my house, my paradise

the construction of the ideal domestic universe

Gingko Press Inc.
5768 Paradise Drive, Suite J, Corte Madera, CA 94925

my house, my paradise

the construction of the ideal domestic universe

Gingko Press
Corte Madera, California
1999

I wish to thank Carmen Hernández Bordas
and Mònica Gili for their untiring efforts in locating the material included in this book.
I am also grateful for the inavaluable contributions made by:

Miquel Adrià
S.C. Albert
Carol Bahnsen
Achim Bunz
Oriol Clos
Roberto Collovà
Savin Couelle
Ted Degener
Claus Dymling
Jordi Falgàs / Fundació Gala-Salvador Dalí
Gustau Gili Torra
Karekin Goekjian
Fernando Gómez Aguilera / Fundación César Manrique
Maricela González Cruz Manjarrez
David Graham
Ian Hamilton Finlay
DeRoy Hurst Jensen / Hearts Castle
International Themes Parks
Shozo Kagoshima / Winchester Mystery House
Irmgard Killing / Bayerische Verwaltung der Staatlichen Schlösser
Cornelius Kolig
Michael Line / California Polytechnic State University
Roger Manley
Toni Nicolini
Louise Noelle Mereles / Instituto de Investigaciones Estéticas
Daniel Paul
Franca Peluchetti / Fondazione del Vittoriale degli Italiani
Denise Pereira da Silva/Quinta da Regaleira
Matteo Piazza
Clovis Prévost
Seymour Rosen
Prince Victor Emmanuel of Savoy
Julia Schulz-Dornburg
Mark Sloan
Alain Tillier
Francesco Venturi

Graphic design: Estudi Coma
Translation: Graham Thomson
Assistant documentalist: Jan Güell Rotllan

GINGKO PRESS Inc.
5768 Paradise Drive, Suite J, Corte Madera, CA 94925
Phone (415) 924-9615, Fax (415) 924-9608
email: gingko@linex.com

Published in the United States of America, 1999

1-58423-015-0

© Editorial Gustavo Gili, SA, Barcelona, 1999

Printed in China

The construction of the ideal domestic universe

House and home

The house is the place where people situate their lives in order to make a home. And it is necessary to differentiate between the concept 'house' and the concept 'home'. The teaching and the practice of architecture inevitably centre on the expressive dimensions of the discipline —the designing of houses— and disregard the notion of home. As architects, we are concerned with the space, the order, the structure, the colour, the light, etc., and leave to one side that more subtle layer, those more diffuse and emotional aspects which are embodied in the term home. There are two quite different personalities involved here: that of the architect and that of the inhabitant. For the inhabitant, the important thing is the capacity to offer accommodation —to bestow significance on a dwelling in the world. The question is: can the 'home' be regarded as an architectural expression? Perhaps we are dealing with a notion that is closer to psychology, to psychoanalysis and to sociology than to architecture.

The home is the individualized dwelling, an expression of an individual's personality and way of life. This subtle personalization seems to lie outside of the scope of our notion of architecture. The home is a complex and diffuse condition, made up of memories, images, desires and fears, past and present; it implies a series of rituals, personal rhythms and everyday routines; it is the reflection of its inhabitant, of their dreams, their hopes, their tragedies or their memories. The setting in which we spend our days is our self-portrait in three dimensions. As the French poet Noel Arnaud tells us, "I am the space where I am", or as Adolf Loos noted, "Your home will be made with you and you with your home". It is neither a building nor an object. The home is an intrapsychological and multidimensional experience which it is difficult to describe objectively. Inhabitation involves psyche and soul, as well as quantifiable formal qualities. It seems clear that the experience of home consists of and includes an incredible range of mental dimensions.

Consider the strange melancholy we feel at the sight of an abandoned house or a demolished block of flats, where the traces and the marks of private lives are exposed to the public gaze. We can not pass by with indifference a house in ruins or burned down. The intensity of such experiences comes not from the fact of imagining the house as it used to be, but because we perceive the home, the life and miracles of its occupants.[1]

The essence of the home, its function as mirror and medium of its inhabitants' psyches is represented more frequently in poetry, in fiction, in films and in painting than in architecture. For example, the novels of Stendhal and Balzac from the early 19th century recount the existence of certain moral ties that unite people with the objects of their interior space. These objects were regarded as indicators of moral character, directly measuring the integrity of the owner. However, "correspondences" of this kind were eclipsed by the consolidation of capitalism in the middle of the 19th century and by the triumph of the self-made man. The interior ceased to be a moral barometer and became instead a domain of material display. The case of the soul was equivalent to the quantity and variety of objects that could be collected and accumulated and, as Marx observed, "the more you have, the more you will be". As an example of the extent to which 'the home' is a psychological concept which it is difficult for architecture to engage with, it is interesting to recall a French experience

Interior views of the apartments in the Edificio Mitre by the architect Barba Corsini, Barcelona, 1963. (© Martí Català).

of the late 19th century. During the last two decades of the century, the interior space was interpreted in terms of a new and specific body of knowledge: the *psychologie nouvelle*. This discipline was promoted by Dr Jean-Marie Charcot, of the Hospital de la Salpetrière, who investigated the interior of the human organism as a sensitive nervous mechanism, with a susceptibility to hypnosis, suggestion and the visual projection of dreams. Although the theory did not have the explosive character of Freud's discoveries relating the sexual instinct with the unconscious, Charcot's researches gave rise to a redefinition of the links between space and being. The transformation of the inner world into a subjective room formed by dreams was not the creation of the artistic avant-garde but was posited by neuropsychiatry. Charcot insisted on demonstrating that suggestibilty and imaginary exteriorization was restricted to patients suffering from nervous imbalances.

Continuing with his theories, and carrying out numerous experiments, Charcot went so far as to interpret the great works of art of the past as documents of clinical pathologies. Rather than expressions of artistic creativity, Charcot argued that the contortions of the statues of Michelangelo, the grimaces on the faces of the gargoyles of Notre-Dame or the tortuous asymmetries of Rubens were objective manifestations of various nervous illnesses[2].

To locate the origin of every work of art in some mental dysfunction is, without a doubt, an extreme and radical posture, however well argued it may be. Nevertheless, it seems evident that, at least as far as the home is concerned, psychology is an instrument of special relevance for our understanding and interpretation of it.

The inhabitant, an active subject

The inhabitant, sane or crazy, is the person who will situate her or his life in the house in order to create a home. In the definition and construction of every domestic space there intervenes a team representing a variety of different professions. The process from the gestation to the materialization of the final product is often long and multidisciplinary. Each of the various professionals involved (from the developer, the architect or the interior designer to the builder, the bricklayer, the carpenter, the plumber, the painter or the estate agent) will have their own particular role assigned to them. Nevertheless, we should bear in mind that the objective, the ultimate point of all of this orchestration of experiences and efforts, is the future inhabitant. But the inhabitant is not only the final client —for whom the domestic space is constructed and who makes its existence possible— but is also, and above all, the person who will take possession of the house, will manipulate it, will utilize it to adapt it to her or his way of life and will endow it with meaning.

There are different degrees of involvement of the inhabitant in the process of definition of his or her own domestic space. From the inhabitant who will buy a standard product (constructed for a standard citizen), to the one who will commission a special product, made to measure (from a team of professionals responsible for translating his ideas), or the inhabitant who will take on tasks normally carried out by the various professionals and will assume functions for which he is not properly trained or educated in the conventional sense of the term. As we shall see in due

Mapamundi Ebstorf. Discovered in 1830 in the Benedictine monastery in the German town of Ebstorf, south of Hamburg, and attributed to the English monk Gervasius of Tilbury, working at the behest of the Emperor Frederick II. The Garden of Eden, represented by Adam, Eve and the serpent, was situated in Asia.

course, these are instances of intrusive architecture or of intrusive construction which only certain people can take on.

Nevertheless, in each of the first three cases, the inhabitant will define his own world, to a greater or lesser degree, and will come to be an active subject within the process of definition of his surroundings. For example, let us consider the case of least involvement, that of the inhabitant of a standard flat in a standard building. If we examine in detail the apartments, apparently all the same within the same building, we will find that in reality they do not resemble one another at all in spite of the fact that they all have the same physical limits, even the same orientation, where it seems that the only objective variant is the height in relation to the street. The real differentiation is the result of the personalizing of the space, not only in terms of the obvious distribution of different objects but in the variation of the spatial quality perceived in each apartment. We might take light as an example. Natural daylight is filtered more or less by different kinds of curtains, which can be single or double, more or less opaque, netted or not. Artificial light will not only be highly variable in terms of the number of lux units assigned to each corner, but also in the quality of the light source. Where one inhabitant will have chosen a standard lamp with a warm shade, another will have opted for a fluorescent strip on the ceiling. It would be easy to enumerate a long list of differences, nuances, exceptions and variants which would serve to reiterate the actively determining role exercised by any new inhabitant on taking possession of their territory. It is clearly a matter of the personal stamp.

We have seen, then, that even in the least degree of involvement, the role of the inhabitant is decisive in the shaping of their domestic universe. So it will be no surprise that this should be magnified to unexpected extremes in all of those cases where the inhabitant decides to take part in the definition of his environment from the very first phase of the process, carrying out in person the work normally entrusted to professionals. We need only think of all of the individuals featured in this book —nonconformist, self-taught, daring, visionary and determined— who decided to undertake in their own right and at their own risk the fashioning of their domestic ideal on the basis of their own personal convictions, infringing all of the established conventions and converting this task into the work of a whole lifetime. These are extreme cases where the only limits are those of the individual personality.

This passion for making oneself the creator of one's own environment is, allowing for differing degrees of intensity, something that is intrinsic to the human species. Think, for example, of the children who play at building a hut in the woods, or hide under the table, making this their own little den. In some sense, we are all born architects. In our contemporary culture, with its high level of specialization, it is refreshing and revitalizing to bring to light a whole world of such infiltrations into the professional domain. In the creation of the domestic space there is, and has been over the centuries, a potential that exists outside of the discipline of architecture; a potential which deserves our consideration at least. In a situation where architectural culture comes very close at times to endogamy, where the latest house by the latest star is disseminated and consumed with genuine voracity all over the planet, it can only be healthy to look at other worlds.

Adam and Eve in the earthly Paradise. 15th. Woodcut.

Paradise

As we have seen, the house is a particular place, concrete, personal and private. This "personal" redoubt is, at the same time, the domain in which the inhabitant has full authority to give form to his or her conception of the world. "A home in the real world is, among other things, a way of keeping the world out"[3]. Inside our own four walls, we are the masters and mistresses. In our own realm we take on the status of "creator gods", free to make our own laws and establish our own world within the world.

On the one hand, the domestic space makes reference to a close and human dimension and, on the other, to a cosmological, divine vision. It is our little-big world. It is our paradise. Perhaps we can not change the world, but once we have recognized and accepted our own limits, the world is ours. Any space imbued with human experience can be regarded as a microcosm, a finite circle of latent energies that invoke a greater whole.

The primitive myth of the Garden of Eden deals with a fundamental sense of loss. Ever since the expulsion of Adam and Eve from Paradise, the fact of being on the outside, locked out of that place, has been a source of perpetual yearning. Return is the dream of Adam and Eve's descendants.

In Paradise, the first men and women lived in perfect peace. They knew neither death nor disease, nor degrading toil nor social imperatives, and they were free of sexual oppression and subjugating powers. What we find in the story of Paradise in the Book of Genesis (the book of origins) is, first and foremost, humanity's mythical dream of a life in harmony with nature.

This age-old dream of the human species was soon to receive an image and a name: *pairi-dae'za*. In the Avestan language of ancient Iran it simply meant an enclosed space or park, without indicating anything of the form of this isolated enclosure protected by a wall. From ancient Iranian to ancient Greek *paradeisos*, it became converted into 'paradise', which can be translated as 'garden of delights'.

The Biblical narrative of Paradise describes, symbolically, a life that is primitive and therefore natural, which at some point in its first beginnings (in any case, before the dawn of history) must have been possible for human beings. For the end of time, a kingdom of total harmony is promised. The ancient reference to a better world in line with the Judaeo-Christian model —and similar myths can be found in many other cultures— is concerned with a natural age, long since disappeared, which serves to designate a promise of future bliss in a beautiful counterpoint.

The decisive element in the story of the Garden of Eden is not the factual veracity of its existence, nor the form it might have had nor where it was situated, but its condition as an inexhaustible model, which has survived as myth by means of literature. We are dealing, in effect, with humanity's age-old dream of a paradisical life, free from cares, abounding in health and an ever greater happiness, and not with the meagre formal description of some historical garden situated in ancient Persia, Mesopotamia or Anatolia.

More than utility, protection or food, medicine or provisions, what this walled precinct guaranteed was a smiling Muse. What was excluded from this garden were the

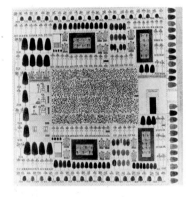

Plan of a garden in Thebes showing the typical configuration of an Ancient Egyptian walled garden (British Museum, London).

dusty earth, the thorns and the thistles of the world of work. The garden has always been a remote and longed-for island; it is distant both in time and in space. In this way fantasy, in the absence of concrete historical details and free of the wearying situations of reality, can develop in freedom. Mythology and art have created many such paradises, examples of which would include Elysium, the Hesperides or Arcadia; islands for which we feel a special predilection, such as Atlantis, the Islands of the Blest, Ithaca, Orplid, Avalon, Cythera, Sicily, Capri, Mauritius or Tahiti; and also the subtropical forest, the Land of Jauja; even America, the Swiss Alps and finally, galactic space have all been converted into this kind of distant and Utopian territory. From the primitive myths to science-fiction, these artificial paradises have been not so much specific places capable of being identified on the map as worlds of fantasy and of dream. But no world of fantasy, no island however fabulous it might seem, is without some basis in reality.

So it is that the dream of paradise contains within it privacy and Utopia, enthusiasm and the escape to the idyllic, the rejection of reality and, at the same time, a breaking away, a desire for change and the yearning for freedom. From this beautiful dream there comes strength, fantasy becomes power and human freedom triumphs precisely there where, otherwise, it would have been obliged to submit to harsh necessity. If we follow the trajectory of such imaginary realms down through time, both pessimistic dystopias and optimistic Utopias, both those which evoke a past Golden Age and the ones projected forwards into a distant future, we clearly see in them the constant urge for political change and revolution[4]. The motivation behind any flight or return to paradise reflects a nonconformist, transgressive attitude to reality.

The Inferno is another imaginary place that has nourished different versions of escapism. Situated at the opposite pole from Paradise, it represents the other end of the spectrum. And, as is so often the case, the extremes merge into one another. The representation of Paradise tends to be accompanied, especially in Romanesque and Byzantine art, by the representation of the Inferno and of the Day of Judgement. This is also the case in a number of the works presented here. The presence of horror, of fear or of the grotesque alerts us to the possible dangers.

Representations of the ideal world: typologies

Paradise is, then, a concept, an ideal place whose definition does not make its form explicit. When the representations of Paradise take shape they do so on the basis of certain models, of certain clichés, of certain architectural archetypes. In pursuing a possible typological classification we encounter time and again the myths of the primitive cave or the grotto, the house-city or the house-state, the Ivory Tower or the Tower of Babel and the garden or the *hortus conclusus.*

The primitive cave

If Paradise belongs to an age that predates written history, the primitive cave takes us back to precisely this antediluvian era. The interior caverns of the Facteur Cheval's Palais Ideal, the grotto in Ludwig II of Bavaria's Linderhof, the volcanic caves of the houses of O'Gorman and Manrique, the Grotto of Aeneas and Dido in Hamilton Finlay's Little Sparta or the troglodyte architecture of the Savoia house are all manifest embodiments of this myth.

A. Kircher. Representation of ancient Mesopotamia, from *Noah's Ark*, 1675. Six walled cities, among which are Nineveh (top centre) and Babylon (bottom centre) with the celebrated spiral tower outside the walls.

The house-city

While the etymological root of the word 'paradise' indicates only that this was an enclosed, walled precinct, it suggests in effect a protected and self-sufficient space. The house presents itself as a house-city or a house-state, regulated by its own laws. Its interior is not only provided with everything that might be necessary but is very often organized as an urban space, with its "streets and public buildings". The name Little Sparta which Ian Hamilton Finlay finally decided to give to his paradise in Scotland, together with the drawing up of his Five Year Plan for the Hellenizing of the Garden are highly illustrative examples of the condition of autonomous republic which these territories can assume. Another example, Il Vittoriale of Gabrielle d'Annunzio, contains its own mausoleum, open-air theatre and chapel (although physically constructed after D'Annunzio's death, they were present in his original conception), all conveniently situated to give a representative quality to its "urban" spaces. We suggested above that what paradise guarantees, in addition to utility, protection or food, medicine or provisions, is a pleasing Muse. This is exactly the case, for example, in the Désert de Retz of Monsieur de Monville. This was equipped with its own gardens to supply the kitchen, with greenhouses for cultivating flowering plants and with its own orchestra to delight the guests at the musical evenings. But what really strikes us most forcefully here is that air of artificial happiness, with its deployment of architectures from so many different geographical regions and ages: the Tartar tent located on the eponymous *Île du Bonheur* (isle of Happiness), the Chinese pavilion, the classical column, the ruined Gothic church, the ice house in the form of a pyramid, etc., offering a different idyllic corner for each day, in which one might, for example, restore oneself with a light al fresco lunch. Similarly, the Linderhof of Ludwig II of Bavaria is composed of numerous buildings scattered across a Romantic Alpine landscape, which transport us simultaneously to the world of the Teutons, to the kingdoms of Arabia, to the Ottoman empire and to the France of the Bourbons. The Bottle Village of Grandma Prisbrey includes, among other elements, a *School House*, a *Little Chapel*, a *Leaning Tower of Pisa*, a *Pencil House*, a *Shell House* and *Cleopatra's bedroom*, all laid out behind the wall constructed from bottles to shut out the dust and smell of the neighbouring turkey farm (we might recall here that paradise effectively excludes the dust of the earth and the thorns and thistles of the world of work).

These are, in other words, worlds in miniature which offer to satisfy all our physical and spiritual wants, so that we will never need to visit the world outside, the real world. It is a matter, then, of creating one's own fiction in order to convert it into a personal reality.

The Tower

Paradise may be either earthly or celestial. In the latter case, the Tower is the ideal construction for elevating us to the realm of the angels. The Tour Eben-Ezer rises up from deep dark underground galleries to the wings of the cherubs that crown it. The Watts Towers defy the laws of statics, as if they were a swarm of wires twisting around one another in order to rise as high as possible. Le Palais Idéal presents ascending routes leading up to its belvedere-terraces, charged with meaning. The

"He had his bed in a cavern."
An illustration from *Journey to the Centre of the Earth* by Jules Verne.

Infinity Room in The House on the Rock is a suspended, weightless space projecting out into the void, at once a cloud and an eagle's eyrie from which can be seen the treetops of the enormous forests round about. The Tower allows us to assume a dominant position that lifts us far from the level of the earth, of everyday reality, in order to situate us above both good and evil. The Tower is one of the most widely utilized architectural archetypes in the history of Utopias, from the Tower of Babel in early antiquity to Tatlin's Monument to the Third International.

The garden

If paradise is "humanity's dream of a life in mythical harmony with nature", the Fenced garden, considered as the *hortus conclusus*, is the most direct model, and the one most often used to represent it. This does not mean that it is any the less rich in nuances. The orchard is the Platonic expression of the garden. Although the trees are nature, the model is not nature but a cultural product with a long history. The siting of the ideal domestic universe in the midst of an artificial nature is the most accurate translation of paradise. The Sacro Bosco de Bomarzo with its grotesque and enigmatic figures, the Paradise Garden of Howard Finster replete with apocalyptic messages, Edward James' La Conchita with its surreal concrete structures, or Ian Hamilton Finlay's Little Sparta with its countless inscriptions reinterpreting the neoclassical garden: all of these are different versions of the same model: the garden.

The inferno

Finally, if the representation of paradise also tends to be accompanied by the representation of the inferno, as we find, for example, in painting, let us see how this is materialized in the construction of the ideal domestic universe. The sins of pride, envy, gluttony and lust are patently evident in the caricatured figures of the Tiger Balm Gardens or in the pot of boiling octopuses at the entrance to The House on the Rock. The Winchester Mansion is a real "house of terror", built not as an entertainment space in a theme park but as a "mousetrap" for the phantoms that haunted its owner. The grotesque mythological figures which surround the Villa Palagonia are monsters that create an atmosphere of unbearable tension in the interior.

Terror is also present, although in a less obvious form, in many other works. In the Désert de Retz, for example, the dimensions of the ruined column indicate that it must have belonged in its day to a colossal temple 400 feet high. A temple on that scale can only be imagined in a land of giants, where we, as humans, are reduced to the size of diminutive Lilliputians about to be crushed under Gulliver's boot.

Just as in neoclassical gardens, in Little Sparta there is an abundance of military iconography. Hamilton Finlay has exquisitely combined "the pleasing Muse of paradise with the terrible devil of the inferno". A small sculpture representing an aircraft-carrier evokes World War II, while at the same time it serves as a bird-table where in winter the sparrows that come down to eat the bread crumbs become inoffensive little aeroplanes. In the idyllic landscape of Lochan Eck there emerges from the water's surface the *Nuclear Sail*, a stone replica of the periscope of a nuclear submarine, a symbol of the enormous menace that lurks beneath the surface.

Mechanisms of the *mise en scène*

In every one of the models described above we find a series of repetitive mecha-

A. Kircher. Representation of the earthly Paradise, situated in Mesopotamia, from *Noah's Ark*, 1675.

nisms for the mise en scène of the ideal domestic universe.

On the one hand it is important to note the continual presence of changes of scale, with either gigantic or miniaturized elements. Thus, as we have already seen, the column in Le Désert de Retz is endowed with colossal proportions; similarly, three gigantic Sphinxes guard the entrance to Le Palais Idéal, as if it were the pyramid of a Pharaoh. The dovecote at Salvador Dalí's house is crowned with an enormous egg that presides over the surrounding landscape, and a pair of hugely over-sized grenades overlook the *Hypothetical Gateway to an Academy of Mars* in Little Sparta. Le Palais Idéal presents on its facades visualizations of various exotic buildings reduced to scale models, while Grandma Prisbrey constructed a miniaturized version of the Tower of Pisa. In every case, the objective of this mechanism is always to transgress reality, to blur the boundaries between reality and fiction.

On the other hand, a whole series of spatial and temporal discontinuities effectively decontextualize the object, and, in the same way as occurs with the alteration of its proper dimensions, carry us off physically and corporeally to the illusory world of fantasy. Thus a Hindu temple is placed next to a Swiss chalet and a mediaeval castle on the west facade of Le Palais Idéal, with the same harmony as existed among the different animals on Noah's Ark.

The construction of the ideal domestic universe is the construction of a fable in which the verisimilitude, the materiality and the intensity of the works cause us to forget that we are dealing with a world of fantasy and thus to accept this as our reality, at least in relative terms and for the length of time we remain inside it.

The protagonists

Behind all of these fables there is always an exceptional individual who performs the double function of creator of the stage-set and chief protagonist. The characters and the backgrounds of these personalities may be extremely varied, quite apart from the material resources at their disposal; their attitudes may be cultured or naive, refined or vulgar; their experiences may be mystical or hedonistic. Whatever the case, these are uncommon characters —idiosyncratic, tenacious, visionary, non-conformist, transgressive, marginal, perhaps eccentric, crazy or extravagant to some people. But always and above all else they are individuals who have, in creating a home for themselves, carried their belief in their own way of seeing the world through to its ultimate consequences.

This belief in themselves gives these individuals the necessary strength to be able to make a reality of their desires (a process that is normally long and by no means free of difficulties), and has its origins in a boundless faith, whether religious or secular. Where that faith is religious, it will not be orthodox. Such individuals have too strong a character to adopt a creed without personalizing it. Both the Facteur Cheval and Raymond Isidore came from church-going families and were brought up in the strictest catechismal orthodoxy, but where religion is present in their work it will have undergone a profound manipulation, and may even be mixed with pagan elements. Robert Garcet, who constructed his Tour Eben-Ezer as a three-dimensional model of his apocalyptic theories, said of himself: "I am not a mystic, I am not a

Hieronymus Bosch, *Garden of Earthly Delights*, left panel, after 1500 (Museo del Prado, Madrid).

believer. So many crimes have their origins in belief! A Bible-reader, yes, a thinker". Howard Finster was a strange preacher who proclaimed himself "a second Noah who has come to this world before it is too late". His work grew out of the need to communicate the message of redemption that God had entrusted him with.

Every one of these house-builders could have pronounced the phrase "where there's a will there's a way". They each achieved their particular dream on the basis of their own, widely different, material realities. In some cases these were individuals with access to very substantial fortunes. Ludwig II "squandered" the revenues of Bavaria in the construction of his castles. Sarah Winchester had an income of $1,000 a day (1900s value) which in spite of all her efforts she did not manage to spend on the construction of her mansion; Randolph Hearst was the head of the greatest media empire in the United States; the brothers Aw Boon Haw had made their fortune from selling tiger balm; Edward James was the son of a wealthy family, etc. In contrast, others had scarcely enough to make ends meet. The Facteur Cheval was a poor day-labourer who was subsequently employed as a humble country postman. Howard Finster repaired clocks, lawnmowers and other bits of mechanical apparatus to raise the money for Paradise Garden. Grandma Prisbrey brought up seven children with-out the help of a husband or regular employment: nothing was going to stop her from building her personal paradise. The degree of direct involvement of each of these individuals also needs to be made clear. Although some of them had qualified professionals working for them, others had to do everything with their own hands and with whatever scrap materials they were able to obtain to achieve their object.

Where professionals were used to carry out certain parts of the work, we should bear in mind that these were merely interpreters. The distinctive imprint is that of the inhabitant, not of the architect or the constructor. Ludwig II commissioned numerous projects from a whole retinue of artists, designers, architects, landscapers and so on, but in the end it was he who wrote, in his own hand, "approved/not approved" on the innumerable sketches which were presented to him before he found the one that fully satisfied his desires. It is perfectly evident that if Linderhof is what it is today, it is thanks to Ludwig. The architect Julia Morgan worked for Hearst for more than twenty years and yet, without dismissing her contribution, San Simeon is the result of Hearst's own designs. The Bergamo architect Maroni was working to D'Annunzio's instructions when he planned the author's house-city of Il Vittoriale. Salvador Dalí also had to employ a local builder, primarily to resolve the potential structural problems posed by his —at times irrational— proposals and construct the house without interfering in the life and work of the great genius. The Villa Spies, that UFO landed on the earth, is more the dream of the extravagant owner, whose life revolved around tourism and aviation, than of its architect Staffan Berglund, however futuristic he might have been. In short, the contribution of the professionals can have only secondary importance.

At the same time, all of these instances of architecture without architects reflect the impulses of a developer-occupier who, free of professional prejudices and theoretical postulates, as has been seen on numerous occasions, could act with liberty and spontaneity.

Hieronymus Bosch, *Garden of Earthly Delights*, central panel, after 1500 (Museo del Prado, Madrid).

Maison aux trois pignons,
drawing by Jean Mar, ca. 1908.
Jean Mar spent more than forty
years as a patient in a psychiatric
hospital, where he devoted his
energies to producing innumerable
 written texts, drawings and small
objects. (From the catalogue of
the exhibition organized by Jean
Dubuffet, *L'Art Brut*, Musée des
Arts Décoratifs, Paris, 1967)

These '*architectes sauvages*' did not set out deliberately to situate themselves in a marginal position. Their ingenuity and spontaneity derive from their purity. The Facteur Cheval picked up stones along the route he travelled each day in carrying out his work. Simon Rodia took advantage of the techniques and materials of the factory in which he was employed. There is a difference between such an attitude and that of people who place themselves in a similarly marginal position by a deliberate act of will, as the hippies of the sixties did. The latter are specifically rejecting the dominant cultural and social order; as anti-consumerists they make use of the waste products of the consumer society, they regard do-it-yourself and build-it-yourself as the best way of protecting the ecosystem, and they isolate themselves on their communes as the ecologically responsible men and women they are. The architectes sauvages know nothing of the science of ecology. They "build it themselves" because they have no other means at their disposal, and use stones, seashells, bottles or stickers because these are freely available. They are transgressors of reality on their own account, not as members of some alternative group or sect. The phenomenon of collective isolation from or transgression of mainstream reality (from the Amish to the punk movement) reflects a particular order of which one becomes part by virtue of obedience to certain rules or codes that leave little space for the expression of the individual personality. The '*architectes sauvages*' were much admired by the artists associated with the Art Brut movement of the middle of the 20th century who sought to focus the attention of art on the work of children, of "primitive" societies and psychotics. To quote Dubuffet: "In order to see at closer range these little works which we tend in general to dismiss. They strike us as rudimentary, vulgar, So be it. But they also transmit in a more immediate way the movements of the spirit and give a warmer, rawer quality to the mechanisms of thought (and not only those of thought). They make use of means that seem preposterous, inadmissible. But we need to consider this: no more than others, perhaps, and we should not allow ourselves to be blinded by habit. Of course their means are scandalously simple, really too obvious. So simple that none of our professional scholars had thought of them! Or they are means that everybody utilizes, but approached in a different way.

"In most cases the artists are people with no academic training, who devote themselves to the work as the occasion permits, who execute it in their own way and to their own taste, without looking to any grandiose destination, driven only by the need to externalize what is taking place inside themselves. A modest art! and one that is often ignorant of what art is."[5]

These architects who have acted on their own account have demostrated that structural invention really has no need of pristine materials: they have happily incorporated all kinds of junk and surplus, available free of cost, into their construction initiatives and their tried and tested technical solutions, erecting structures of considerable size and complexity. As a result, their "outsider" architecture, which we might define as an approach to construction that owes nothing to established techniques, has very often come into conflict with local planning regulations and disgruntled neighbours. [6]

All of these works, self-built or otherwise, have evolved organically, as Dalí observed: "Our house has grown exactly like a genuine biological structure, by cellular addition". The need to create the ideal domestic universe emerges here as a vital necessity. And thus, as a vital necessity, while there is life there is work. These are evolving achievements that have advanced episodically, by stages, each new addition and modification coinciding with a particular period of their creator's life. In some cases the process continued every day, day in, day out, year after year. These are genuine works in progress. In very few cases did the procedure involve the drawing-up in advance of plans or sketches of the whole. Instead, it progressed from the vision to the creation by impulses. In this context, it is important to appreciate that while these projects are endless they are not on that account incomplete. Although Edward James' La Conchita has an apparently unfinished look, it would be absurd to dismiss it as an incomplete work simply because there was no specific programme, or planning, or project.

From private to public: from personal fantasy to collective fantasy
If we have noted that these are works created in order to enclose a private world of fiction, and are in that sense introverted, we also need to consider their contacts with the outside world. On many occasions they are also the scenario in which we would like to situate ourselves and, as such, were conceived for the purpose of show and presentation. Il Vittoriale was the setting in which D'Annunzio received his visitors, and no matter how important those visitors were, they had to come to D'Annunzio if they wished to see him. Thus Mussolini himself made the journey to Lake Garda especially to meet the hero of the nation. Port Lligat started out as an ascetic painter's refuge, to become in due course the scenario chosen by Dalí for his much-publicized public life. Hearst constructed his castle as a fortress, but also as a hotel in which to receive his many guests. Grandma Prisbrey ran guided tours of her Bottle Village at 25 cents a head; she explained her work to the visitors and, at the end of the tour, entertained them with a short recital on the piano. Howard Finster created his Paradise Garden in order to spread the holy message of his work to the world. The Tiger Balm Gardens put forward a moral and cultural lesson and at the same time advertised a commercial product. Alex Jordan converted his House on the Rock into a genuine amusement park, open to the public. In other words, the self-referential private universe opens up, little by little, in order to allow others to share in its creator's achievement.

This gradual opening-up of the private precinct to the public gaze culminates after the death of the creator-inhabitant. A review of the present situation of the majority of the houses featured here seems to suggest that this is the only valid option. If the person —both the means and the end— by and for which they were created no longer exists, the only remaining possibility is to open it up as a place of homage to the individual and his or her work. These universes are too personalized to be a home to anybody other than their original owners.

This movement from the private to the public may offer us a useful clue to understanding the present-day situation with regard to the construction of the domestic paradise. Although there are still today (and there will continue to be as long as there

Hubert Robert, *The swing* (Metropolitan Museum of Art, New York).

are human beings) those *raras avis* who build their own worlds of personal fiction with the same intensity, it is evident that the model which currently predominates is a totally different one. The domestic paradise that is being constructed now in certain parts of the world is a programmed domestic paradise. Far from being the brainchild of its creator-inhabitant, this tends nowadays to be an artificial domestic paradise which reflects the dream of a social group, which is not built but bought. We might think here, for example, of the residential developments which the Disney Corporation has laid out in Florida. Celebration is a suburban town, planned as the original 1955 Main Street of the Disneyland in California, but this one has real people living in it. A series of neotraditional buildings recreates a perfect fantasy replica of small-town America, with its wooden porches, picket fences and old-fashioned details. Here, too, we are living in a dream space, but one that is hardly our own, given that a company has taken care of everything, so that we only have to organize the removal in order to take our place in paradise. A place where even the colour of the houses and the layout of the gardens are regulated by contract and supervised by agents of the property company does not allow much room for individual freedom of expression.

In general, the artificial is now a collective artificial. Not only in the domestic sphere but in every other area. Theme parks proliferate all over the planet. In Beijing a world in miniature is being constructed, so that the Chinese will be able to visit different countries and regions without leaving home, just as Ludwig did at Linderhof. In Japan they are building an artificial beach where the sun will shine 24 hours a day, just as hot as the real sun but with the harmful rays screened out, and the salinity of the water will be controlled so as not to sting people's eyes. Here the waves will be artificial, like the waves in the lake of blue waters —of the same blue as the water in the Grotta Azurra on Capri— at Linderhof, which was also equipped with the machinery to produce waves or to create a rainbow exactly according to Ludwig's wishes. If, as we have seen above, paradise was formerly the product of individual self-expression, nowadays it is just another commercial leisure facility. This is the contemporary paradise, to which we gain admittance by buying a ticket at the cash desk. It would seem that we are witnessing a democratization or a commercialization of paradise, depending on how we choose to look at it.

Miniature from the *Roman de la Rose, ca.* 1485. (British Library, London).

1. JUHANI PALLASMAA, *Phenomenology of Home. The new private realm*, The Berlage cahiers 3, 010 Publishers, Rotterdam, 1995.

2. DEBORA SILVERMAN, *A Fin de Siècle Interior and the Psyche. The Soul Box of Dr. Jean-Martin Charcot*, Daidalos no. 28, Seelen-Kisten / Soul Boxes. Berlin, 1988.

3. JOHN SEABROOK "Home on the Net", *The New Yorker*, October 1995: "A home in the real world is, among other things, a way of keeping the world out".

4. CARL FRIEDRICH SCHRÖER, *Architecture of Gardens in Europe*, Benedikt Taschen Verlag GmbH, 1992.

5. JEAN DUBUFFET, *L'Homme du commun à l'ouvrage*, Éditions Gallimard, Paris, 1973.

6. ROGER CARDINAL, "Clarence Schnidt, Arquitecto Intruso", *Fisuras del mundo contemporáneo* n. 3/4, Madrid, 1997.

Following page. Universal CityWalk, Universal Studios, California. The Jerde Partnership International. (© Stephen Simpson).

Sacro Bosco di Bomarzo Pier Francesco *Vicino* Orsini, Prince of Orsini, Viterbo, Italy, 1552

The Sacro Bosco is a garden situated at the foot of the hill of Bomarzo, on which the residence of the Orsini family stands.

In the opinion of his contemporaries, Vicino Orsini was well deserving of respect. He had an exuberant and cheerful nature, an insatiable curiosity, a great sense of humour and a broad humanist culture. He had read contemporary authors such as François Rabelais, Torquato Tasso and Ludovico Ariosto, and was especially interested in books about India, about alchemy, treatises on longevity, the Cabbala and the Romantic epics. He dreamed of travelling in the East.

The prince had known heartache: his first wife, Giulia Farnesio, died young. This tragic note in his biography contains some of the keys to understanding the meaning of the garden.

With the aid of the Neapolitan architect, archaeologist and painter Pirro Ligorio, Vicino Orsini created a garden with an enigmatic sequence of monuments laid out along a path.

The garden, laid out on a slope, had numerous pools and cascades of water. Two Sphinxes, lions with women's faces, flanked the entrance, and it is thought that one of these faces is a likeness of Giulia, Orsini's first wife. There are inscriptions on the pedestals of the Sphinxes, one of which reads: "You who enter here with the idea of seeing all that is within, tell me later if all these wonders have been made by trickery or by art", while the other reads: "Whoever does not walk through this place with eyebrows raised and lips pressed tight, will also be incapable of admiring the famous Seven Wonders of the World". With these affirmations, Orsini made it clear that he had set out to create the eighth wonder of the world.

The visitors, having been thus advised of what lies in store for them, move on to encounter Ligorio's monumental sculptures, left where nature had scattered the rough blocks of stone from which he carved his monuments. Here they will find elephants, dragons and irate giants. Each of these pieces is, by virtue of its own enigmatic iconography or its position in relation to the whole, part of an initiatory route. For example, we will see a tortoise bearing a globe on its back, on top of which balances an ideally beautiful maiden with her arms raised, advancing slowly but inexorably towards the gaping jaws of an enormous whale.

The interpretation of the series of pieces has given rise to a variety of different readings, some considering it as representing Orsini's military past, others proposing interpretations based on mythlogy or literature. It is not easy to arrive at any definitive conclusion, but of note among the wealth of possible readings of the Bosco Sacro is the one based on the *Hypnerotomachia Poliphili* (The dream of Polyphilus), regarded as the most beautiful and enigmatic book of the Italian Renaissance.[1] In terms of its development, the book follows the scheme of an amorous romance in which Polyphilus relates to his beloved the dream that, in five stages, will lead to their union with one another. The principal intention of the work is to represent the culture of antiquity (the architecture, the gardens, the art and the dress), not confining itself to a mere description of the classical models but seeking to transcend these through the exercise of the imagination. Both book and garden take the form of a narrative, a love story which is in turn an allegory of spiritual progression.

1. KRETZULESCO-QUARANTA, EMANUELA. *Les Jardins du Sogne. Poliphile et la Mystique de la Renaissance*, Societé d'Editions Les Belles Lettres, Paris, 1996.

The mouth of an enormous monster at the top of a flight of stairs with the inscription *OGNI PENSIER VOLA* (All thought flies), ready to swallow up the visitor.

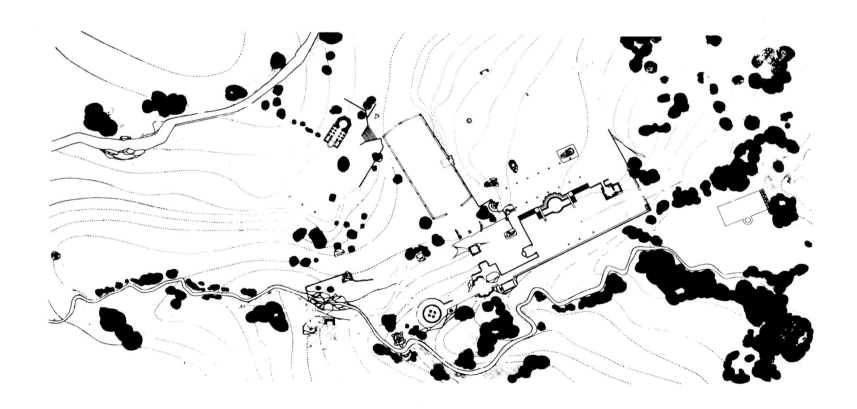

General plan and section

Elephant with a tower on its back, recalling both the elephant of Polyphilus with the obelisk, and the countless Indian elephants used to carry howdahs. Here the elephant is carrying a tower, the *Turris Sapientae*. The elephant is lifting up a neophyte in its trunk.

A giant tortoise with a globe on its back, on top of which balances an ideally beautiful maiden with her arms uplifted.

The Tower of Meditation.
A house with a square plan,
probably used in its day for
reading and meditation, fre-
quented by the cardinals who
were friends of *Vicino* Orsini.

A sculptural group representing a mermaid and a pair of lions symbolizing the fertility of the irrigated earth; an enormous reclining giant; one of the Sphinxes which guarded the entrance to the garden, with enigmatic inscriptions on its pedestal, and Ceres, the Roman goddess of agriculture.

Villa Palagonia The Prince of Palagonia, Bagheria, Sicily, Italy, 1746-1792

In the 18th century, Bagheria was the fashionable location of choice where the wealthy burghers and the aristocracy of nearby Palermo had their splendid country villas. The most remarkable of these is the Villa Palagonia, nowadays surrounded by the tall buildings of a dense, chaotic suburb.

The villa was built around 1715 by the Dominican architect Tommaso Maria Napoli —no other works by whom are known— for Francesco Ferdinando Gravina, prince of Palagonia. The third owner, his grandson, remodelled the villa to transform it into an eccentric baroque caprice, filling the house and the garden with statues of monsters and curiosities of all kinds, which only a mind such as his could imagine.

The villa stands at the centre of an estate marked out by a series of circular routes. All around its perimeter there are auxiliary constructions which make up a thick wall that encloses the property and insulates it from the world outside. Originally, access was by way of a broad avenue, also flanked by high walls, culminating in the patio or front garden. The villa, with its rectangular plan so distorted as to be curving, and with its corners almost touching the boundaries of the property, forming four different and independent exterior patios. The volume of the villa is set back to create numerous terraces and recesses. The most important of these is the one which accommodates the flight of steps of the principal access. This is a stairway of several flights which turns and doubles back and eventually makes its tortuous way to the main door. It is precisely this sensation of conflict or tension that the prince wished to exalt in his reform of the villa. Along the avenue giving access there are strange figures on pedestals, on top of the perimeter wall and around the house. These statues, sculpted from the soft local stone, are varied in their repertoire, with an abundance of mythological figures out of context, imaginary beasts, dwarfs, harlequins, busts of men on top of women's or fishes' torsos, and a variety of bizarre mutations, such as a human face with an elephant's trunk instead of a nose. The figures standing on the walls are silhouetted against the sky and look towards the interior of the property. The sources of inspiration for this iconography can be traced, in part, to the prince's interest in the Middle and Far East, as well as to the literature of the period, although all of this has been "suitably" exaggerated or manipulated.

In the interior, after passing through an oval vestibule, on one side there is a grand mirrored salon, and on the other a series of bedrooms. The salon contains a series of busts on its marble walls representing the prince's ancestors. The ceiling is covered with small faceted mirrors which follow its vaulted form, mirrors which must have reflected in a thousand and one fragments the image of the social life that was conducted in the salon. This significant vision is at the same time a reflection of the character of the villa's owner and his marginal, not to say strange, relations with the social class to which he belonged. As for the furniture, which has long since disappeared, we are indebted for our information to the reports of a number of illustrious visitors of the period. As Goethe remarks in his *Travels in Italy*: "The chairs have lost some of their legs, so that it is impossible to sit on them, and if any one looks as if it might be a place of repose, the butler warns us of the dangerous nails concealed beneath the satin upholstery"; and "on the tables there were exotic pagodas, Chinese and Japanese figures. Perhaps the most surprising thing was the columns made of coffee cups, sauce boats, bowls, jugs and teapots all heaped up in utter confusion, including some pieces of fine porcelain".

The Villa Palagonia is a clear exponent of the way the obsessions, the schizophrenia or the psychosis of the owner can come to be channelled into domestic architecture. According to local legend, the prince was physically deformed, marked by malaria, and already advanced in years when he inherited the villa, and married a beautiful young woman only 16 years old, Maria Gioachina Gaetani, who was not without other admirers. Consumed with jealousy, the prince remodelled the villa as he did precisely in order to drive these rivals away from his wife.

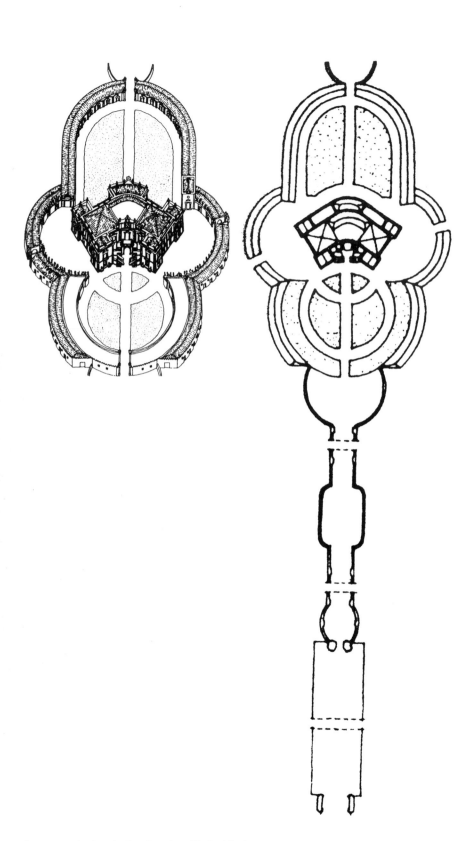

Axonometric sketch showing the villa inside the
walled precinct, with its curving lines, and plan with
the original avenue giving access to the villa. View of
the sculptures flanking one of the doors.

The stairs leading up to the *piano nobile* with their series of flights turning and doubling back in one of the recessed spaces of the villa before arriving at the main door.

Strange mythological figures:
imaginary beasts, dwarfs,
harlequins, busts of men on
women's or fishes' torsos and
a human face with an
elephant's trunk are among
the sculptures silhouetted on
top of the perimeter walls.

Le Désert de Retz Monsieur de Monville, Retz, Chambourcy, France, 1774-1789

Le Désert was constructed for François Nicolas Henri Racine de Monville. This picturesque garden of 38 hectares was situated on the outskirts of Paris, close to the royal estates of Versailles, Rambouillet and Marly. The garden contained a total of 20 folies, or *fabriques*, as they were called in France at that time. These constructions were strategically sited in order to offer artfully composed landscaped views, the most important of them being the *Colonne Détruite* (ruined column) in which Monsieur de Monville made his home. This column, 23 metres in diameter and 25 metres in height, had five floors, one of them below ground, thus inverting the hierarchy of scale by carrying the concept of the false ruin to surreal extremes.

Monsieur de Monville was a gentleman close to the court of Louis XVI whose fortune it was to live through the end of an era —the *ancien régime*— and a revolution. De Monville came from a family of wealthy financiers; his bachelor estate was briefly interrupted by a single, short-lived marriage: his wife died only five years after they were married. He then devoted himself to the worldly life, for which he was eminently qualified, thanks not only to his elegant appearance —testified to by his innumerable female admirers of the time— but to his talent. He was an accomplished dancer, and was never without invitations to the most sumptuous balls; he played the harp with great skill, he was an expert horseman and an excellent shot with the bow, with which he hunted pheasants in the nearby Bois de Boulogne, to widespread admiration. Although his sybaritic and hedonist character was criticized by some, with the result that he was never offered the diplomatic posts to which he had aspired as a young man, he did move in aristocratic circles, becoming intimately acquainted with the "little" rustic shepherd's refuges where the private passions of royalty found their release. And he reproduced the character of those *maisons de rendez-vous* very well when he came to construct his own dream.

The ruined column is a stone construction in which a series of rooms —oval, circular, semioval or semicircular— were laid out around a central cylindrical staircase, lit by a skylight at the top from which hung baskets with flowering plants, selected from the ones cultivated in the greenhouses. The residual spaces were used to accommodate less important rooms or as corridors. The square, rectangular and oval windows were reflected in succession in the mirrors placed on the adjacent fireplaces, offering different views as one walked around the circle, from window to window. The top floor was lit from above, making it possible to have a solid wall on the facade in order to simulate the irregular section of a real ruined column.

An inventory of the contents of the property would include an entrance which simulated a grotto on its rear side; the "ruins" of a classical temple dedicated to Pan; the "ruins" of a Gothic church incorporating fragments of an original 13th-century church in Retz; a Chinese house with a French-style interior which was Monsieur de Monville's first residence before the column was completed; a stable, to complete the pastoral effect; greenhouses; vegetable gardens; the *Île du Bonheur* (isle of happiness) with a painted wooden construction designed to resemble a tartar tent; a cottage; a tomb; a pyramid, based on the Roman mausoleum of Caius Cestius on the Appian Way, used to store the ice brought from the Alps; an obelisk; an open-air theatre; a neoclassical construction known as the *Temple of Repose*; a rustic bridge; two artificial lakes; a small altar and various other auxiliary constructions. In short, everything necessary to create a world of one's own.

With regard to the design of the complex, although it seems clear that the owner himself exercised overall control, other hypotheses have suggested different names: Boullée, who had designed and built two houses for de Monville in Paris; François Barbier, a student who took de Monville to court to obtain payment for his work as draughtsman, or Hubert Robert, the well-known landscape painter. There have even been speculations that the programme was drawn up on the basis of Masonic ritual, although there is no evidence that de Monville was a member of that fraternity.

Ground floor, second floor, third floor and attic plans, section and elevation of the ruined column, from engravings by Le Rouge.
(Georges Le Rouge, "Jardins anglo-chinois", Cahier XIII, Paris, 1785. *Nouveaux jardins à la mode*, published by Georges Louis Le Rouge between 1776 and 1787, is the most important illustrated book on the garden design of the 18th century. One whole portfolio was devoted to the Désert de Retz. Some of the illustrations are slightly different from the built reality.)

Watercolour showing the column. The silhouette is more dramatic and the column is wider than in Le Rouge's drawings. Attributed to Racine de Monville, *ca*. 1785.

The oval window on the third floor
and view of the column following the
reconstruction carried out by the
architect Olivier Choppin de Janvry
from 1972 until 1993. Above: members
of the Frédéric Passy family, who
owned the Désert from 1856 until 1936.
They lived in the column, and
introduced a number of alterations such
as the regularization of the silhouette
and the creation of new windows; view
of the interior from this period.

View of the ruined Gothic church
and engraving by Le Rouge; views
of the *Temple of Pan*.
Below: detail of the kitchen
garden where the fruit and
vegetables were grown, making the
house virtually self-sufficient.

The Chinese house in a state of abandonment. It stood beside the lake until 1912, when it collapsed.

View of the Tartar tent on the Isle of happiness and engraving by Le Rouge.

The pyramid used to store ice.

General plan of the Désert after Le Rouge. The key does not indicate the Tartar tent, the *Temple of Repose*, the rustic bridge or the little altar, although these constructions were illustrated in other engravings.

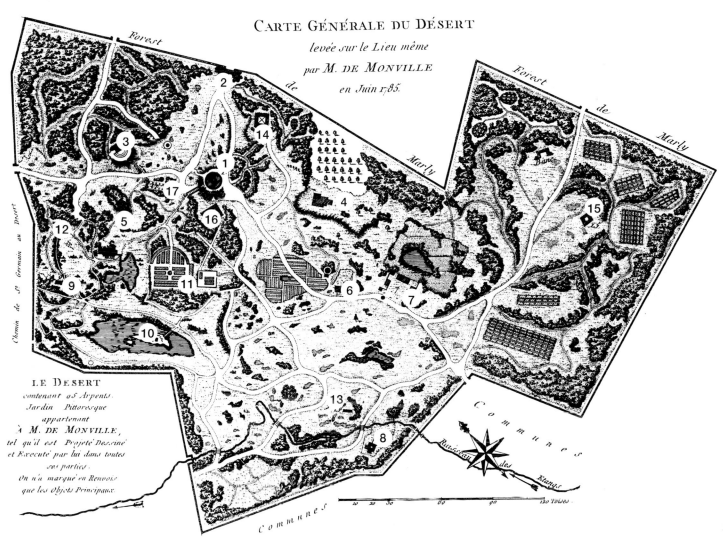

CARTE GÉNÉRALE DU DÉSERT

levée sur le Lieu même

par M. DE MONVILLE

en Juin 1785.

LE DÉSERT
contenant 95 Arpents.
Jardin Pittoresque
appartenant
À M. DE MONVILLE,
tel qu'il est Projeté Dessiné
et Executé par lui dans toutes
ses parties.
On n'a marqué en Renvois
que les Objets Principaux.

1	Ruined column	10	*Isle of happiness*
2	Grotto at the entrance to the garden	11	Greenhouses
3	*Temple of Pan*	12	Cabin
4	Ruined Gothic church	13	Tomb
5	Chinese house	14	Pyramid
6	Dairy	15	Obelisk
7	Porter's lodge	16	Servants' quarters
8	*Hermitage*	17	Open-air theatre
9	*Orangerie*		

André Breton and his surrealist group in front of the Saint Germain entrance of the Désert de Retz, *ca.* 1960.

Linderhof Ludwig II of Bavaria, Linderhof, Germany, 1869-1879

Linderhof is one of the three castles built for Ludwig II of Bavaria. Ludwig came to the throne at the age of 18; he never married, and had no children, and he found the affairs of state of relatively little interest. His real obsession was building castles and recreating the spaces of Wagner's operas, or more precisely, Wagnerian themes. A Romantic figure par excellence, retiring, solitary and antisocial, Ludwig II has been extensively studied. He devoted his short life to the construction of his own private world, far removed from the everyday reality of the country he ruled over. A tormented spirit, he closed the doors on the outside world and sought fulfilment in his fantasy realm, constructing and decorating three-dimensional sets. The historical figures he admired most were Louis XIV and Marie Antoinette. After a trip to Paris in 1867, on which he visited Versailles, he decided to convert the rustic hunting lodge his father owned at Linderhof into a replica of the park of Marly. Here he designed a French garden with sharply defined geometries, in contrast to the bucolic Alpine landscape in which it was set. This was the first of a long series of decontextualizations. Linderhof was transformed into a microcosm, with buildings from many different periods and places.

The castle itself is an eclectic construction in the baroque-rococo style, with ten reception rooms and only one bedroom (there was no need for more, since guests did not figure in Ludwig's programme). The relationship with the surrounding landscape was studied with great care. Not only were the views of the castle from the exterior subject to elaborate control; Ludwig had watercolours painted of the images to be seen through each of the windows, precisely determining the configuration of these exterior spaces with their fountains, cascades, sculptures and pools.

Ludwig used to get up at 5 o'clock in the afternoon; then the fountains were turned on, he would go out onto the terrace, salute the bust of Marie Antoinette, take a stroll, look at the newspapers, preferably illustrated, have his second breakfast, discuss policy with his advisers —especially how to get rid of the social democrats— and then, at night, he would often go for a sleigh-ride in the dark. In the early hours of the morning, on his own, he would stop at the grotto. The sources of inspiration for this were the Grotta Azurra of Capri and the Grotto of Venus in Tannhäuser. Apparently carved out of the living rock, this grotto was actually built with a vaulted brick roof, and adorned with stalactites and stalagmites fashioned from iron, canvas, mortar and plaster. An irregularly shaped artificial lake was the scenario for sailing, in a boat in the form of a swan. Ludwig was prepared to employ pioneering technology to achieve the desired atmosphere. The first dynamos in Germany were used in the grotto at Linderhof, to power the lamps that created special effects of light and colour, mechanisms to produce artificial waves on the lake and a device which could recreate rainbows.

Another of the keys to understanding this microcosm is the Hundinghütte (Hunding's cabin), a replica of the set of the first act of *Die Walküre*. This was a rustic cabin of wood, with a great tree-trunk presiding over the interior. Ludwig himself set out to select the right tree. When he finally found the perfect trunk, it turned out to be a beech, when the stage directions of the opera called for an oak. The problem was solved by having the beech clad with oak wood. The log fire and the bearskin rug for sleeping on constituted the *attrezzo* recreating the atmosphere of the ancient Teutons.

Ludwig's fondness for oriental architecture led him to buy an Arab kiosk (actually constructed in Berlin) at the International Exposition in Paris in 1867. This was duly transported to Linderhof, where its colours were changed to adapt it to its new environment, and a throne in the form of a peacock installed in it. On a subsequent trip to Paris Ludwig bought another building, a Moroccan house, which was dismantled for transportation, and duly reassembled and extended. At Sachen, at the top of the mountain, Ludwig built the Königshaus (the king's house), a hybrid in the form of a Swiss chalet with a Turkish salon in the interior.

Ludwig immersed himself completely in his own theatre, where he was not a spectator but the leading actor.

View of the hall of mirrors in the interior of the palace.

View of the south-west facade of the palace, Baroque-Rococo in style; view from the main balcony with a perspective comprising several stretches of water, of note among these being a pond with a fountain; the music room decorated with pastoral scenes and the bedroom, situated at the geometrical centre of the palace, between the lilac room and the pink room.

Overall plan of the park of Linderhof, from the project drawn up by Carl von Effners in 1874.

The Arab kiosk purchased by Ludwig II at the International Exposition in Paris in 1867. The throne in the form of a peacock, the marble fountain and the windows lit from the exterior are good examples of the adaptation carried out by Ludwig II.

Following pages: the construction of the grotto was commenced even before certain parts of the palace, since it was one of the principal objectives in Linderhof. Ludwig used to sail on the lake in a swan-shaped boat. It proved far from easy to obtain the desired colour of blue for the grotto. To Ludwig, it was of the greatest importance that it be exactly the shade he had decided on. He sent artists to Capri to study the intensity of the Grotta Azzurra, he ordered numerous tests of the light and consulted several chemists with a view to altering the components of the pigments to obtain the desired effects.

Plan of the grotto in 1877 and drawing
by Heinrich Breling from 1881.

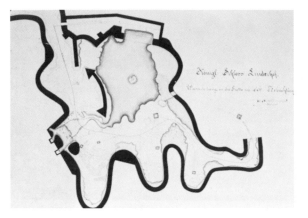

Studies of the Grotta Azzurra on
Capri by Christian Jank, and coral
furniture.

Hundinghütte. The original cabin burned down, and it has been rebuilt on several occasions. At the time of the most recent reconstruction, in 1990, its position was changed for reasons relating to the line of the border with Austria. On the following page: the hunting lodge which Ludwig II built in Sachen. In fact, he took very little interest in hunting. The real motive behind this construction was his fascination with the sublime Alpine landscape.

The Königshaus at Sachen is a hybrid construction. Two very different worlds were brought together in the interior of a single building: a rustically austere Alpine interior and an exotic and highly elaborate Turkish salon.

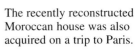

The recently reconstructed
Moroccan house was also
acquired on a trip to Paris.

Le Palais Idéal Facteur Cheval, Hauterives, France, 1879-1912

Joseph Ferdinand Cheval was born in 1836 at Charmes-sur-Herbasse, in the north of the Drôme region. At that time the inland area of the Drôme was rural, eminently agricultural, marked by the poverty resulting from a series of floods and epidemics. Cheval's first job was as a day-labourer, which during that period paid what was virtually a starvation wage. Nevertheless, neither these privations, nor the premature death of several members of his family, were sufficient to extinguish the energetic, willing, lively, tireless and tenacious character of the man who was later to recall: "As I walked along I thought of Napoleon I, who said that the word impossible ought not to exist. Hands full of the future..." At the age of 30 Cheval decided to become a postman. He was allocated a rural round of 32 kilometres, which he travelled on foot. These long daily circuits gave him the opportunity for lengthy meditatations, fusions with the landscape and intense moments of solitude. A mystic and visionary, he received his revelation ten years later. As he describes it, on the significant day that was to change his life radically, he tripped over a stone sculpted and eroded by the weather which opened up his mind to new ideas. His fascination for the stones he found along his way can be seen as the starting point of an obsession that was to stay with him for the next 33 years: the idea of building his own ideal palace.

The Palais Idéal has no concrete functional programme. It is an uninhabitable monument, sublimely useless, but essential in so far as it constitutes the expression of an obsession. It is a grand assemblage of *objets trouvés*, where fossils, sea-shells and stones eroded by the water are brought together to create a landscape of soft and fluid forms. In terms of the built complex as a whole, the four facades are of outstanding importance, above all if we bear in mind the limited dimensions of the interior space. This interior consists of no more than a few small grottoes or caves which, by means of labyrinthine passages and twisting stairways, lead us towards the upper belvedere-terraces. Each facade is a complex independent creation in its own right, although considered as a composite whole, they reveal the possible existence of a meaning to the itinerary which would situate Le Palais Idéal somewhere between the origin of the world and eternity. An initiatory progression defined by an invisible circle conducts us from the north facade (the phase of life) to the west facade (the phase of death) where, by way of a series of interior passages, we are brought to the upper terraces where (after revisiting the west and south facades) the whole experience culminates on the east facade (the phase of life reincarnate). These fossils, these eroded stones and these shells are mementos of the origin of the world. All of the cavities are places of rest, of a return to the subterranean life and to primitive times.

The iconography employed here is very varied. The sources drawn from are extremely diverse, with a great abundance of fusions of historical figures and a decontextualization of objects. A number of different far-away countries (India, China, Egypt, etc.) and architectural archetypes (the Hindu temple, the Swiss chalet, the Arab mosque, the medieval castle, etc.) are reinterpreted, manipulated and combined according to the singular vision of the creator of Le Palais. There can be no doubt that the postman had a very rich imagination, although it is probable that it was, in part at least, nourished by the magazines of the period and their coverage of the Universal Expositions in Paris.

There are continual changes of scale, sometimes miniature, sometimes gigantic. The hypertrophy of the figures, as well as the constant spatial and temporal breaks, give an unexpectedly grotesque quality to the whole. Numerous quotations, names, proverbs or popular sayings are inscribed in different corners, combining to form a text loaded with visionary messages.

The Facteur Cheval, initially criticized by his neighbours, gradually attracted the curiosity and the admiration of foreign visitors, to the point of becoming the most paradigm case in the recent history of spontaneous architecture.

Jean Pierre Jouve/Claude & Clovis Prévost. *Le Palais Idéal du Facteur Cheval*, A.R.I.E. Editions, Hédouville, 1994.

The only known plan drawn up
by Joseph-Ferdinand Cheval,
consisting of six drawings; the
west facade and the south facade
in 1905.

Sous la garde des TROIS
GEANTS j'ai placé
L'EPOPEE des HUMBLES
courbes sous le Sillon

Socrate

Sous la garde des TROIS
GEANTS j'ai place de
L'EPOPEE des HUMBLES

Winchester Mystery House Sarah L. Winchester. San Jose, California, USA, 1884-1922

Sarah Lockwood Pardee was born in 1839 in New Haven (Connecticut). In 1862 she married William Wirt Winchester, son of the owner of the patent of the "rifle that won the West". The couple's only daughter died soon after her birth, in 1866. When tuberculosis took her husband in 1881, the 42 year-old widow inherited a fortune of 20 million dollars and 48.8 per cent of the shares in the Winchester Repeating Arms Company, which provided her with an income of $1,000 a day.

The widow consulted a clairvoyante in Boston, who attributed the untimely deaths of her daughter and husband to the vengeance of the spirits of the people killed by the rifles manufactured by the Winchester family. To escape the curse, she was told, she would have to go out west, buy a new house and undertake a continuous process of construction, in accordance with the directions given by the spirits. In this way she would be able to free herself from them and perhaps find the key to eternal life. In 1884, Sarah Winchester bought an 8-room farmhouse in San Jose. Up until her death, 38 years later, the work went on seven days a week, 24 hours a day. She employed between 18 and 20 servants, between 10 and 22 carpenters and between 12 and 18 gardeners. She invested more than five and a half million dollars in the house, transforming it into a Victorian mansion with a highly unusual functional programme: 160 rooms, 47 chimneys, 40 stairways, 19 kitchens, 2,000 doors, 10,000 windows and 13 baths. The number 13 is repeated obsessively here: 13 hangers in the wardrobes, candelabras for 13 candles, stairways of 13 steps, 13 windows in a room, walls and ceilings of 13 panels each, and many more. There were thirteen sets of gold cutlery on the table where she dined every night in company with 12 invisible guests.

Each night she slept in a different bedroom, but between midnight and two in the morning she went to the blue room of the spirits where she received instructions from the other world and planned the work for the following day. This room is right in the centre of the house, with no windows, with one way in and three doors leading out, one of which opens onto the void space just above the kitchen. In order to disorient the spirits, she built stairways that lead straight up to the roof, skylights projecting up from the ground, doors that open onto solid walls, locked doors at the end of labyrinthine passages and corridors with down stairways followed by up stairways, configuring apparently absurd geometries.

The construction was seriously affected by the great earthquake of 1906, which demolished three of the seven floors and changed the principal axis of the house's growth from the vertical to the horizontal.

The house is mid-Victorian in style, with vaguely Oriental details, and is elaborately ornamented. The built volume is the result of the agglomeration of all kinds of architectural elements. Turrets, cornices, windows, chimneys, balconies, porches, curving and straight walls, arches and cupolas are found in a variety of positions, sizes and finishes.

Sarah Winchester embraced the idea of progress and the new construction technologies in the building of her house. She was a pioneer in the installation of an automatic lift on the west coast, in the use of mineral wool for insulation and in the production of gas for the lamps in the house and grounds, which lit up at the touch of a button. She patented a laundry for washing the clothes and invented a system of latches for the windows based on parts of the Winchester rifle.

The house's 65-hectare Victorian garden has a great variety of trees and plants, laid out in geometrical patterns, as well as a number of sculptures and a shrine. Around the perimeter, a thick, high hedge protects the property from prying eyes.

Sarah Winchester lived in solitude, establishing a minimal relationship with two of her employees, and keeping her face hidden behind a veil at all times.

Exterior view of the mansion and the ballroom, probably never used, with its floors and walls decorated with six different kinds of wood. Above: the bedroom where Sarah Winchester died in 1922, and detail of the roofs. Below: corridor and view of the storeroom where she accumulated precious objects, many of which were never actually installed. On the following page: the little door measuring only 120 centimetres high which opens onto a dismountable landing.

View of one of the lanterns which
let light into the floor below and
allowed the lady of the house to
keep an eye on her servants, and
view of a set of false wardrobe
doors which in fact conceal
openings which provide a line
of sight through a series of 30
consecutive rooms.

Staircase consisting of 44 steps with risers only 6 centimetres high, 7 complete turns and a horizontal length of more than 30 metres, in spite of having a vertical ascent of only 2.75 metres, and view of one of the flights of stairs leading nowhere. Above: the Y-shaped staircase which enabled the servants to move rapidly between the three main levels of the house.

Quinta da Regaleira. António Casrvalho Monteiro, Sintra, Portugal. *ca.* 1892-1910
Sintra —with its delightful and healthy microclimate and its singular geographic position on the far western edge of the continent of Europe, on the shores of the great
Atlantic ocean— is a place of idiosyncratic characteristics which has been enthused
over by foreign visitors since the late 18th century. Lord Byron, for example, went so
far as to describe it as a Glorious Eden. This setting was very much in keeping with the
ideals of a certain type of society —bourgeois/liberal/Romanic— wich flourished
during the 19th century, and Sintra soon found itself in vogue. To have a house in Sintra
was a sign of wealth, of good taste, of political and social importance. As a result, little
palaces and chalets —all of them marked by an eclectic taste for the revival of past styles— began to spring up around the urban centre and spread across the slopes of the sie-
rra. This was the context in which António Carvalho Monteiro purchased the Quinta da
Regaleira in 1892. The famous 'Monteiro of the Millions', the nickname given to this
enigmatic figure, born in Río de Janeiro in 1848, the son of Portuguese parents who
brought the young Monteiro back to Portugal to receive a decent education. The heir to
a considerable family fortune, which he multiplied in Brazil thanks to the monopoly of
the trade in coffee and precious stones, Monteiro was a cultivated man, a distinguished
collector and bibliophile, whose greatest passions were books, the opera, musical instruments, clocks, sea-shells, butterflies and antiquities. A philanthropist and patron of
the arts, he spent large sums on the construction of his fantasy. A conservative, a monar-
chist and a Christian, he developed his own system of beliefs, particularly in religion,
where he espoused a gnostic Christianity in line with the theological doctrines which
guided the early Christians and were later followed by the Knights of the Order of the
Temple. There is no documentary evidence for his supposed involvement in initiatory
sects, but it can be inferred from the nature of a number of the books in his library and
from the members of his small circle of friends that he had at the very least a considerable knowledge of esoteric lore. The Quinta da Regaleira constitutes an initiatory itinerary which asks to be understood in terms of alchemy, freemasonry, the Portuguese
mythological tradition and the cult of the Holy Ghost.
When Monteiro resolved to construct his philosophical mansion, he enlisted the aid of
Luigi Manini —an architect, painter and set designer of Italian origin— to make his
dream a reality. The Quinta da Regaleira is an estate of almost four hectares wich a
series of buildings laid out among splendid gardens. A variety of plants and trees from
different parts of the world alternate with a sequence of stairways, landings, belvederes,
galleries and terraces, the whole conceived as if it were the set design for a theatrical
performance. In 1898 Monteiro turned his attention to the residence, the so-called
Renaissance House, a castellated building near the palace. After remodelling the coach-
house and stables, and above all from 1904 on with the inauguration of the Chapel dedicated to the cult of the Holy Trinity and the palace in 1905, he commenced a period of
intense construction employing a large team of artisans. The Chapel of the Holy Trinity
pursues the same decorative line as the palace. A great number and diversity of crosses
decorate the church, some of them Latin, others Maltese, for the Templars, yet others
palaeo-Christian. Beneath the chapel an austere crypt, in sharp contrast with the delica-
te neo-Manueline adornments of the chapel, evokes the monastic knights of the order
of the Temple. This is like a second chapel underground, pervaded by that gloom in
which men of faith may encounter the path to self-Knowledge and to closer commu-
nion with their ancestors. The subsoil of the estate is burrowed through by a whole
series of galleries, grottoes and caves, most of them artificial, taking advantage of the
geological characteristics of the granitic mass of the Sintra range of hills. Of particular
note among these descents into the underworld is the initiatic well. This has always been
dry, a kind of inverted tower submerged in the depths of the earth, with an unusual door-
way covered by a rolling stone moved by a concealed mechansim which affords entry
to the other world. In flights of fifteen steps the spiral stairway descends by way of nine
landings, supported by numerous columns. At the bottom a great Maltese cross, allied
to an eight-pointed star, constitutes the deepest core of the complex.

Detail of the *Dove Girl* on the
main facade of the palace; figures
of fantastic animals crowning the
building.

DENISE PEREIRA, EDUARDO GEADA, JOAO RODIL, JOSÉ MANUEL ANES, *La Quinta da Regaleira*, Fundaçao Cultursintra, Sintra. 1998.

Hunting hall; facade of the chapel and detail of the Delta in the interior of the chapel.

Plan of the complex
  1 *Initiatory well*
  2 *Grotto of the Virgin*
  3 *Imperfect well*
  4 *Dragon fountain*
  5 *Entrance of the guardians*
  6 *Celestial terrace*
  7 *Lake of the cascade*
  8 *Garden of Perpetua*
  9 *Tower of the Regaleira
      and grotto of Leda*
 10 Aquarium
 11 *Fountain of abundance*
 12 *Egyptian fountain*
 13 *Garden of the camellias*
 14 *Grotto of the cathedral*
 15 *Swan lake*
 16 *Balcony of the gods*
 17 *Palace of the Regaleira*
 18 *Chapel of the Holy Trinity*
 19 Orchid house
 20 *Terrace of the Chimeras*
 21 *Renaissance house*
 22 *Garden of the Ladies*
 23 Incinerator
 24 Coach houses

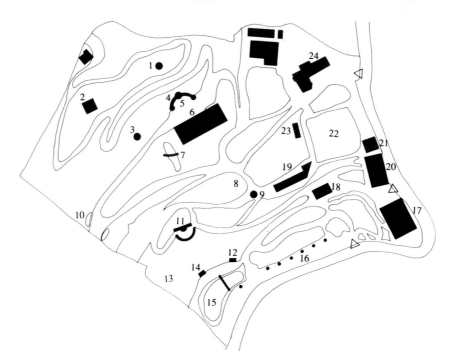

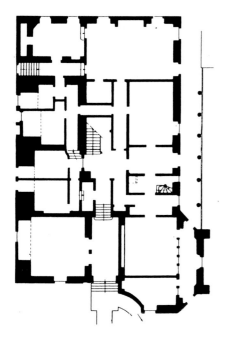

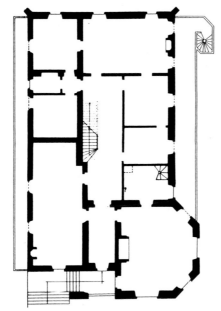

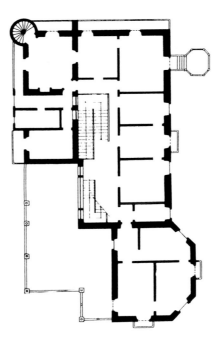

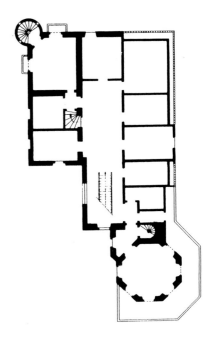

Underground plan   Ground floor plan   First floor plan   Second floor plan

Previous page: *Chapel of the Holy
Trinity* (1904-1910); *Balcony of
the gods*; *Tower of the Regaleira*
(1800-1893) and *Grotto of Leda*.

Initiatory well.

Celestial terrace and *Bench of the "515"*. Below: detail of the guardians at the entrance to the subterranean realm.

Grotto of the lake of the cascade.

Hearst Castle William Randolph Hearst. San Simeon, California. USA 1917-1937
William Randolph Hearst liked to go horse-riding and camping out on his property near San Luís Obispo, California, accompanied by his five children. It reminded him of the excursions he had gone on with his father, George, who had set about extending the ranch as his business interests —gold, silver and copper mines— expanded. Hearst employed the profits from his media empire to take the extension even further. In 1917, at his favourite place on the estate, the highest hill, surrounded by tall cypress trees and with a fantastic view of the Pacific Ocean, he presented to his guests a model of the castle he intended to build right there. The property consisted principally of the Casa Grande, or 'ranch' as Hearst liked to call it, surrounded by other constructions such as the Casa del Mar, the Casa del Monte and the Casa del Sol, an enormous outdoor swimming pool and a private zoo.

Hearst Castle came into being with a double vocation: as a private hotel for his numerous guests and as a place in which to accumulate fragments of architecture and works of art that only a triumphant American magnate could have amassed. Hearst Castle is a mixture of styles, from the Spanish baroque to the Italian Renaissance, without forgetting the classical world of Greece and Rome. Both Hearst and his architect, Julia Morgan, declared on more than one occasion that what they were really constructing was a museum of architecture. Julia Morgan spent two decades working on the house, faithfully incorporating the constant changes of plan imposed by her client.

During the day, Hearst's guests were free to do as they pleased: swim in the outdoor pool of Neptune or in the indoor Roman pool, play tennis or cricket, go horse-riding, fishing or walking. After deciding that the Neptune swimming pool was too small, Hearst had it extended in 1927. The facade of a temple and a portico of classical columns, of white marble inlaid with green, surround the enormous pool of mineral water.

The one fixed appointment of the day was at half past seven in the evening, in the meeting room, where the guests were greeted by Hearst and his companion Marion Davies, with whom he lived for 32 years in spite of the fact that he was still married. Dinner was served at nine in a room known as *The Refectory*, a dining room where the guests sat on 14th-century Catalan choir-stalls, around a long set of 16th-century tables from an Italian monastery. Above their heads, just beneath the vault of a Renaissance monastery, hung 20 standards from the Palio of Siena. The walls were covered with Flemish tapestries from the 16th century. After dinner, the company would move to the projection room, where they would watch a film.

Hearst was a passionate collector. By means of his agents, constantly travelling around Europe, and his own trips, he acquired art objects and architectural elements of all kinds. He had a special passion for elaborate ceilings. His collection included a number of coffered ceilings from historic buildings in Italy and Spain. On not one but two occasions, his agents bought an entire Spanish cloister, which were then dismantled stone by stone and shipped from Spain to Hearst's warehouse in the Bronx. There he stored an enormous quantity of pieces, some of which were eventually sent out to California over the course of the years.

The two towers of the Hearst Castle complex contained the more private rooms and Hearst's offices. A lift of carved mahogany —originally the confession box of a French church— went up to the *Celestial Suite*, where Hearst had his own bedroom. The collapse of Hearst's business empire obliged him to halt the work on the Casa Grande in 1937, and he was forced to sell the house in 1947.

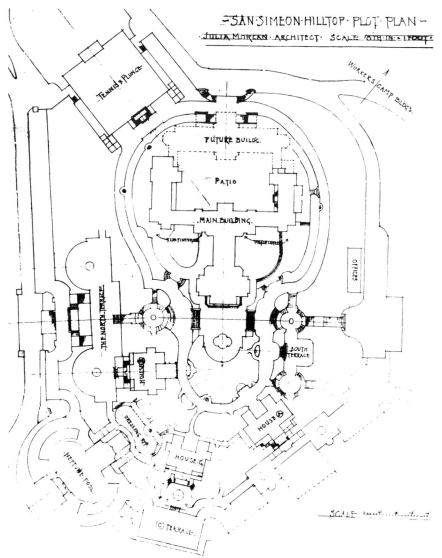

General plan, *ca.* 1930, and aerial view. Above: views of the facades of the Casa Grande and the Casa del Sol.

Interior views of the Casa Grande: *The Refectory*; bedroom of the Doges; meeting room and the Doges' living room. Below: detail of the library.

Outdoor Neptune swimming pool
(1935-1936) and indoor Roman
swimming pool (1927-1932).

Il Vittoriale degli Italiani Gabriele d'Annunzio, Lago di Garda, Italy, 1921-1952

At the age of 52, d'Annunzio went off to fight in the First World War as a pilot and naval commander. In the course of the war he was decorated with various medals for his prowess and bravery in dramatic military episodes, consecrating his position as Italy's great charismatic hero. And at the end of the war, in 1919, for more than a year he led the famous occupation of the town of Fiume, claimed by the Italians but eventually ceded to what was to become Yugoslavia. D'Annunzio's army also served as the military reference for the emerging fascist movement, in everything from their uniforms to their terminology. Benito Mussolini took part in the episode, and subsequently plagiarized d'Annunzio's stage-setting and his form of addressing the multitude in his speeches. The coming to power of Mussolini established d'Annunzio as the national hero, and he was ennobled with the title of Prince of Montenevoso.

In 1921, d'Annunzio bought a villa on the shores of Lake di Garda, which he gradually reformed and extended, effectively converting his residence into a national monument, with the idea that Italy and himself would be inseparably linked for ever more.

The Villa Cargnacco had been expropriated from its German owner during the war. The first reforms consisted, in d'Annunzio's words, in "de-Germanizing" it. He brought in a local architect, Gian Carlo Maroni, essentially because he was a war veteran and an irredentist, and by making a donation to the Italian state he ensured for himself the means necessary for the construction of his private world.

In the same way that the Romantic nationalism of the 19th century and the totalitarian ideology of the fascist state apparently needed a double-jointed contortionist reasoning to establish their association with one another, Il Vittoriale brings together the picturesque intimacy of a miniaturized Italian village with the abstract monumentality of the public buildings.

And just as in d'Annunzio's novels, plays and poems not a single word is spoken that does not have the authority of the illustrious figures of the past, in Il Vittoriale there is no room, piece of furniture, object or stone that does not have some precedent, some source, some model. Il Vittoriale is made up of a sum of copies, doubles, clones manipulated and distorted by D'Annunzio, and in this lies its extraordinary suggestive power.

The entrance to the complex is by way of the Piazza del Vittoriale, where signs and motifs pay tribute to the founder of this city-state, to whom Italy is indebted for poetic treasures and the salvation of the nation. The plaza is presided over by the monument to the fallen, with a triple arcaded porch, next to the church of Sant Nicholas. The route leading to D'Annunzio's residence, La Prioria, is an eclectic, lyrical composition in the manner of a collage of unconnected events which passes through various different spaces in a sequence of streets, squares and gardens.

The facade of La Prioria, overlooking the Piazza Dalmata, has coats of arms and architectural details inserted between a variety of literary excerpts. Each of the rooms was composed with care, as it it were a poem, the reading of which discovers D'Annunzio's nostalgia for times of past glory. Every kind of paraphernalia, from the aeroplane in which he flew over Vienna to a desiccated tortoise, by way of altars, busts, paintings, photographs, figures of animals, musical instruments, weapons and autographs, illustrate the narratives obsessively repeated in the mind of the comandante.

The open-air theatre evokes the role played by political and artistic discussion in the formation of the national spirit, while at the same time reproducing the structure of the amphitheatres of Ancient Greece.

Between tall cypress trees, facing out towards the waters of the lake, is the prow of the *Puglia*, whose captain was killed in the Fiume adventure. After the prow had been dismantled and transported to the lakeside, the rest of the boat was completed in stone.

Representing the mortality of the human in contrast to the eternal survival of the nation, d'Annunzio's remains occupy the centre of the mausoleum, a circular composition made up of the tombs of ten legionaries who died in Fiume (in some cases purely symbolic) and which stands on the highest point of the nine hectares of Il Vittoriale degli Italiani. After d'Annunzio's death, Maroni continued the work in accordance with the instructions communicated to him in séance sessions by the comandante's spirit.

Gabriele d'Annunzio and Benito Mussolini during the Duce's visit to the Vittoriale in 1925, with the SVA 10 aeroplane suspended from the cupola of the auditorium.

General view from above
showing the situation of the
estate on the shore of the lake.
In the foreground is the
mausoleum among cypress trees.

Piazza Dalmata and view of the
wing occupied by the archives
and library.

General plan of the estate
 1 Gardone Riviera Sopra
 2 Monument to the Fallen
 3 Entrance plaza
 4 Monument to Victory
 5 Theatre
 6 Bandstand
 7 Piazza Dalmata
 8 Schiafamondo
 9 *La Prioria*
10 Mausoleum
11 The *Puglia*
12 Arengo
13 *Lake of dances*

Entrance to the Piazza Dalmata and Monument to Victory.

Following pages: open-air theatre, mausoleum and the *Puglia*.

Previous page: *Veranda of the small Apollo*; Zambracca; the dining room and the poet's study with his desk.

The music room.

Watts Towers Simon Rodia, Los Angeles, California, USA, 1921-1954

The son of Italian peasants, Sabato Rodia was born in 1879 near the town of Nola in the region of Campania. Preceded by his older brother, Sabato emigrated to the United States at the age of 15, where the two were employed in the coal mines of Pennsylvania. Sabato changed his first name to Simon, Samuel or Sam, but he refused to give up his family name as his brother had done, who started calling himself Dick Sullivan in order to integrate more easily into the mining community, for the most part of Irish origin. Rodia's reputation for hard work was not enough to enable him to be apprenticed to a skilled trade, since he could neither read nor write Italian and had considerable difficulty in expressing himself in English. When his brother was killed in an accident, he went off to Seattle in search of new opportunities, and there he married his first wife and fathered three children.

Simon Rodia, temperamental and rebellious by nature, never missed an opportunity to voice his disagreement with the members of his family, with government policies on immigrants, with the degrading evolution of the relations between parents and children or with the damage done to the image of his personal historical heroes, among whom were Christopher Columbus and Marco Polo. Rodia was regarded as an anarchist and a drunk, a weakness which probably began after the sudden death of his daughter.

Rodia took certain liberties with regard to his marital status, his ability to read and write, his age and his naturalization as a US citizen. Having divorced his first wife, he married a 16 year-old Mexican girl and went back to California. This second marriage was also short-lived, and four years later he married another Mexican woman, also an immigrant to the USA. He then started work in a tile factory and, with the assistance of one of his brothers, bought a plot of land in the working-class town of Watts. On this triangular plot, at the end of a dead-end street bordered by the train and trolley tracks, Rodia made his dream reality. After his third wife left him, for the first few years Simon Rodia shut himself off from the world with the determination to "do something" ("I am gonna do something", he used to say) in that flat, banal landscape, devoid of meaning, surrounded by orchards and humble single-storey houses. For 33 years Rodia devoted to this enterprise the major part of his modest income and all of his energy, working on his Towers in the evenings and at weekends. His shy, antisocial character, combined with the difficulty he had in expressing himself, prevented him from explaining exactly what he meant by "doing something". When people asked him about the meaning of his sculptures, he used to answer, falsely and cynically, that they were a monument to peace after his experience as a soldier in the First World War (which in fact he had avoided by emigrating to the USA), or a monument to the great country that had welcomed him, or to the freeways of California, or to the place where his wife was buried, or simply that he had stopped drinking.

After Rodia fenced in his plot of land, his neighbours in Watts began to see the silhouettes of various peculiar structures emerging above the fence, of note among these being three tall conical towers pointing up into the sky. In addition to these, which later came to be known as *Niña*, *Pinta* and *Santa María*, he constructed another two smaller towers, the Ship of Marco Polo, a summerhouse, the *Garden Spire*, a barbecue and a fishpond, as well as a pergola and a chimney next to the house. 'Our Village', as Simon Rodia used to say when he referred to his group of constructions, is the product of the ingenuity and effort of one man, built using a special technique he developed with the meagre resources at his disposal. He bent pieces of steel angle sections into the shapes he wanted with the aid of basic tools or lengths of train track, which he used for leverage. Neither bolted nor welded, these pieces were held together with wire mesh and covered with mortar pressed into place by hand. He then added pieces of ceramic, glass and other found materials to the wet surface. Once the structure had hardened, Rodia climbed up it with the necessary materials on his back in order to carry on building higher and higher.

In 1954 he suddenly left the place and went to live in Martinez, California, where he died in 1965.

Plan

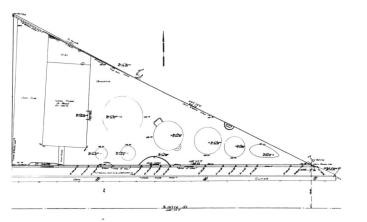

After the plot was fenced in, the Watts residents saw a number of peculiar structures emerging above the fence, of note among these being three tall conical towers rising up into the sky.

In 1959, the Watts Towers were at the centre of a widely publicized controversy about their artistic value and their safety, and the arbitrary application of building regulations to such a singular work, which would have resulted in its demolition. The bureaucrats were opposed by a group of intellectuals and experts, who set up the Committee for Simon Rodia's Towers.

In 1961, when Simon Rodia was shown photos of Gaudí's Sagrada Familia, all he said was: "Did he have helpers? I did it on my own".

La Maison de Picassiette Raymond Isidore, Chartres, France, 1928-1964

Raymond Isidore married, at the age of 24 years, a young widow with three children. In response to his new condition as husband and stepfather, he began the construction of a new home. In 1928 he bought a piece of land on the outskirts of Chartres, behind the cemetery of Saint-Chéron. Although Raymond Isidore found work in a local metal foundry, his real job (to which he devoted the rest of his working life) was as the municipal employee responsible for the upkeep of the cemetery. A man with very deep-rooted religious beliefs, Isidore considered that he had kept faith with his spirit, and stuck to the path he was meant to follow. He visualized this as a narrow road which terminated at the small door that led him into Paradise.

The task of constructing a home for his family grew out of the desire to create a place in his own image. Raymond Isidore was quite adamant that this home would not be thought up by an architect and imposed on the family against their wills.

The construction work commenced with the house itself, but this was to grow and evolve over more than thirty years, always guided by the spirit. Raymond Isidore claimed that something showed him the way to proceed; when he pictured an image in his head, it was somehow transmitted to his hands, to his fingers, and he set to work. He imagined the thing that was to be produced and proceeded to construct it without any preliminary design, modifying and rectifying it by feel, as it took shape, guided at all times by that visionary intuition.

Little pieces of porcelain, marble and ceramic are everywhere. Isidore started with the interior, with the dining room, then the bedroom, and gradually the whole thing overflowed, spreading and taking over the furniture, the sewing machine and the kitchen until it had appropriated the whole of the family space.

He covered the walls of his house with landscapes inspired by postcards (the door opens onto Mont Saint-Michel). His admiration for places of worship, which he considered to be not only beautiful and imposing buildings but also works of great complexity that had to be the fruits of spiritual revelation, prompted him to undertake their repetition and multiplication. Cathedrals and mosques, the temples of Jerusalem and other religious architectures appear on all sides.

In the garden there are two thrones. The *Black throne,* the throne of the ordinary man, the cemetery caretaker, stands in front of a tomb where Isidore used to sit down to rest after he had finished his day's work in order to admire from there his city and its cathedral without leaving the house. The tomb is not in fact Isidore's own, but a black sarcophagus which attests to the material nature of the body in contrast to the spiritual nature of the soul. The *Throne of the spirit,* coloured blue, of course, has no tomb and is positioned opposite a representation of Jerusalem.

La Maison de Picassiette situates us in front of 'the other space', the space that transports us from Paradise lost to Paradise regained.

CLAUDE AND CLOVIS PREVOST, *Les Bâtisseurs de l'imaginaire*, Editions de l'Est, La Malgrange, 1990.

The widow Picassiette in the bedroom and the dining room.

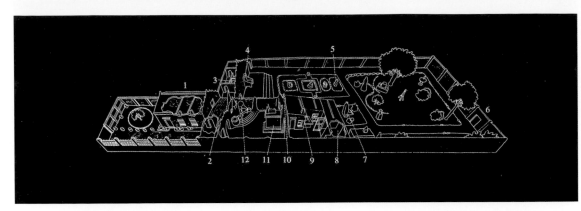

Perspective of the complex
1 The house
2 The chapel
3 The kindergarten
4 *Blue throne of the spirit*
5 *Monument to the fields*
6 The old kitchen garden
7 *Monument to the Virgin*
8 La Gioconda
9 Summer house
10 *Chartres, the city and the cathedral*
11 *Tomb of the body*
12 *Black throne*

View of the interior of the chapel and the bedroom, with the bed, the night table, the vases, the sewing machine and the musical instruments were covered with fragments of pottery and porcelain. Facing page: view of the garden with the *Monument to the fields* in the foreground. Following pages: *Chartres, the city and the cathedral*, and view of *Black throne*.

Casa Salvador Dalí Salvador Dalí, Port Lligat, Cadaqués, Spain, 1930- *ca.* 1971

In 1929, Dalí met Gala, the woman who was to be his companion and muse for the rest of his life. The couple decided to find a home of their own, and Dalí bought a fisherman's cottage in the out-of-the-way location of Port Lligat. Port Lligat is one of the 'most arid, mineral and planetary places on Earth', Dalí declares in *The secret life of Salvador Dalí*. Remote and difficult to get to in those days, it fulfils all of the conditions necessary for an ascetic life, a shelter and the desired isolation. The house itself is a simple shack, 22 metres square, originally used for storing fishing tackle, and in this too it corresponded to Dalí's preferences: "I wanted it to be very small, the smaller it was, the more intrauterine". Like the rest of the constructions that made up the original village of Port Lligat, this is rectangular in plan, with walls of unmortared stone built directly on the rock, between 4 and 5 metres across, and a single-pitch tiled roof.

This shack represents the starting point of a long process by means of which the ascetic painter's refuge was to be converted into the scenario of an 'actor'. We should bear in mind that Dalí was an artist whose private activity was not as important as the complicated scenic expression of his work. [1] The process of transformation was biological and evolutionary. The house, in constant evolution, came to be something continuous, modifiable, always in a state of metamorphosis. It was physically expanded, both by the addition of a second storey to the original volume, and by annexing the neighbouring cottages. In the many times repeated operation of extending the original volume, Dalí always made a point of conserving the existing roof, so that it was possible to live on the lower floor while the work was being carried out, leaving a triangular chamber and in this way maintaining the character of the original spaces, focused on the window. The annexing of the various neighbouring cottages ended up configuring a maze-like complex made up of a series of separate rooms. In the interior, the results of this system of composition is reflected in a complicated scheme with a multiplicity of intersecting sightlines, narrow stairways connecting the different levels of the cottages, and twisting passageways leading from one space to another, creating different sensations of surprise or contrast.

If the physical and spatial framework of the house is in itself irrational, Gala and Dalí were responsible for exacerbating this condition to the limit, through the imposition of an enigmatic layer of significations. A profusion of curious objects of all kinds and origins appears on all sides. These objects are in many cases of little significance in themselves, but take on a special relevance when seen in their context. They are transformed, manipulated, juxtaposed and positioned in the space in such a way as to become surreally expressive. In a manner fundamentally similar to that with which he approached his paintings, Dalí's cottages and objects represent existing matter, which he set out to transform in order to create from it his surreal universe.

In the material realization of his projects, Dalí had the assistance, from 1935 on, of Emili Puignau, a young builder from Cadaqués. At the outbreak of the Spanish Civil War, Dalí and Gala went to live in New York, where they remained until 1948, and construction was interrupted during this long period. On his return, after years leading a worldly life which helped to establish his reputation on the art circuit, building work was restarted in order to satisfy a series of new requirements. Dalí needed a space to work in, a made-to-measure studio, spaces for storing the accumulated pictures and accommodation for the live-in servants. Thus, a further three cottages were duly incorporated into the complex. This phase, in which the bedroom, the library and the living room also assumed their definitive form, culminated in 1953 with the construction of the dovecote as the symbol which dominates the complex and asserts the leading role of the house within the village of Port Lligat.

Dalí carried on adding to the house up until the early seventies. The Vía Láctea (or Milky Way; the white path parallel to the sea), the oval lounge, the outdoor spaces such as the patio, the wall, the summer dining area and the swimming pool were the last additions.

Port Lligat in 1930, with the original fisherman's shack. Portraits of Dalí and Gala at two different stages in their lives: the first, which dates from 1931, in the interior of their ascetic refuge, and the second, from the sixties, in the scenic oval room.

Evolution of the house in nine phases.

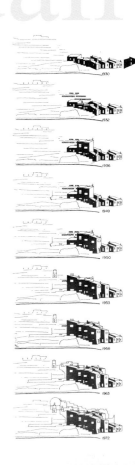

1. ENRIC GRANELL TRÍAS, *Dalí Arquitectura*, Fundació Caixa de Catalunya, Barcelona, 1996.

Sections, ground floor and first
floor plans. (Plans drawn by Oriol
Clos i Costa, architect).

Following page: views of the
library and the entrance hall and
detail of the dining room.

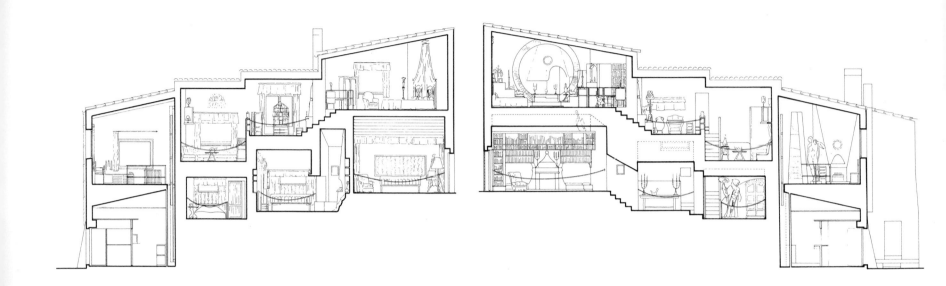

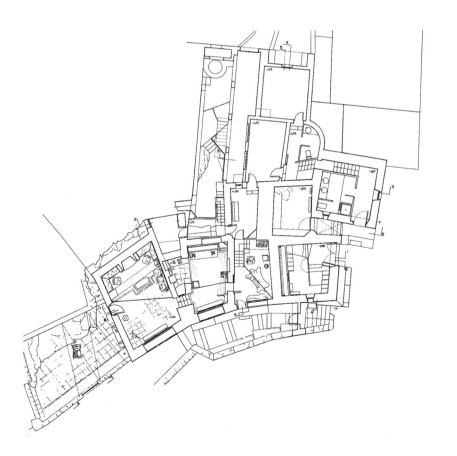

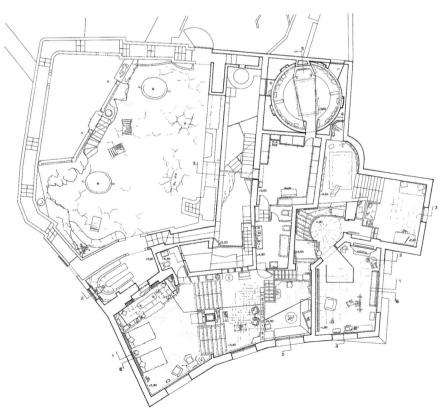

Previous page: view of one of the
flights of stairs, shaded by a large
parasol; view of the studio with the
door in the shape of a
truncated isosceles triangle, and
view of the models' room with walls
and ceiling covered with moiré
paper, creating optical effects.

The *Bird's room* and the main
bedroom. *The Bird's room*, so called
because it contained a cage with
canaries, was the link between the
previous room and the bedroom.

Previous page: the oval room was the last to be built. This was a private, almost inaccessible precinct, which could only be reached after passing through the dressing room and the closet room. One of the few rooms with a door was Gala's private retreat.

The patio, sheltered from the wind, with its enormous white planters in the shape of gigantic cups and curved mill-stone seats. View of the heads of Castor and Pollux which crown the construction.

The area around the swimming pool was the centre of the house's social life. Note the pavilion at one end, its design based on the housing of a radio with the old lens from the nearby Cap de Creus lighthouse in its interior. The benches are positioned to give a view of the bathers. The swimming pool, inspired by the gardens of the Alhambra, features a reproduction of the Fountain of the Lions. This is the part of the house where the influence of American Pop art is most in evidence, as denoted by the advertisements for tyres, the sofa in the form of a gigantic pair of lips or the telephone booth.

Haw Par Villa Tiger Balm Gardens Aw Boon Haw, Hong Kong and Singapore, *ca.* 1935

Aw Boon Haw (the tiger) was only 15 years old when he and his brother Aw Boon Par (the leopard) took over the running of the family business. Their father, a herbalist who practised medicine, had emigrated from the province of Fukien, in the south of China, and finally settled in Rangoon. After his death, his two sons inherited the family business, and his talent for invention. In 1910 the brothers patented an unguent, a mixture of camphor, cloves and senna oil, which became world famous as tiger balm. With a variety of uses, for example, as an analgesic for soothing headaches, rheumatism, insect bites or muscle pains, this product earned them a great deal of money. And as Aw Boon Haw used to say: 'The wealth that comes from society should be used for society'. On the basis of this charitable philosophy, they helped to build schools, hospitals and orphanages. Another of their philanthropic interests was the transmission of Chinese history and culture. Accordingly, when they decided to build themselves a mansion in Hong Kong, they also designed a garden which was open to the public, and then went on to create a second version of this garden in Singapore. These gardens were intended to achieve several objectives at once: to educate and improve the cultural and moral development of the ordinary people and, at the same time, to be a leisure park where the masses could relax and enjoy themselves; in addition, they were also a highly effective advertisement for the brothers' most successful product.

The property in Hong Kong had an area of more than three hectares, on a hill with views of Happy Valley. Although it has since been eclipsed by tall buildings, in its day it was a notable part of the skyline of the city as seen from Victoria Harbour.

In laying out the park, stones were imported from China to fashion the landscape. Six artists from Swatow were specially brought to Hong Kong and Singapore in order to train the teams of local craftsmen. The results were these highly unusual gardens, charged with mythology and symbolism, and often described as an inferno in the form of a landscape populated with nightmares.

The varied repertoire of figures, all of them painted in bright colours, serves to warn the visitor of the terrible punishments in store for wrongdoers. The particular moral vision of the owner makes itself apparent everywhere, with the most graphic details, in scene after scene of grotesque caricatures (traitors cut in half, corrupt officials crushed under enormous weights, prostitutes fried in oil, rapists decapitated, liars tortured and thieves burned at the stake).

These are very far from being classical Chinese gardens, although they do embody Chinese folk culture. The traditional Chinese garden represents the geography of the country, in miniature: mountains, rivers, lakes, vegetation, etc. In Aw Boon Haw's gardens the plants and the landscaping play a purely secondary role. Here the Taoist or Buddhist sphinxes, the monsters, the heroes and villains are the protagonists of a Taoist version of purgatory.

Aw Boon Haw's mansion in Hong Kong, situated close to the entrance to the gardens, is a three-storey building modelled on the classical Chinese pavilion, but constructed of reinforced concrete, by the side of which is a garden of bonsais.

In the Tiger Balm Gardens we once again find a recreation of the primitive cave. A series of tables, stools and ledges which appear to be sculpted out of the rock furnish this rest zone, with its countless stalactites and stalagmites.

La Conchita Edward James, Xilitla, San Luis Potosí, Mexico. *ca.* 1945-1984

Edward James, whose mother was an illegitimate daughter of King Edward VII, was noted in aristocratic circles for his extravagance. After his schooling at Eton, where his favourite subject was literature, he worked for the British diplomatic service. During the twenties he associated with the transgressive artistic avant-garde of surrealism, acquiring an important collection of pictures as a patron of Salvador Dalí, Max Ernst and René Magritte, among others. This period saw the consolidation of the spiritual restlessness that was to mark James' life. Towards the end of the Second World War he was in Los Angeles with Aldous Huxley, studying Buddhism. It is interesting to note that he was one of the group of intellectuals calling for the conservation of the Watts Towers. In the mid forties, during his time as editor of the review *Minotaure*, he took a long trip through Mexico, where he discovered the exuberant charm of a place known as Xilitla, to the south of San Luis Potosí. Here a sign was revealed to him, an initiatory, epiphanic vision of the metamorphosis into a human butterfly of the friend with whom he was bathing in a natural pool. Seduced by the place, and in the hope of shaking off his personal troubles, he commenced construction of La Conchita, a venture which was to occupy him for the rest of his life.

In the midst of the forest, in a place marked out as special by local Indian tradition on the steep slopes of the Espina hill, with its waterfall and its exuberant vegetation, Edward James constructed a series of magical buildings dotted over the 30 hectares of his property. His work, in its excess and prodigality, sets out to establish a dialogue with the nature of the place. While the revolutionary ethic opposed itself to bourgeois morality and its institutions, surrealism sought to transgress the apparent dominant order through the development of an antifunctionalist aesthetic which questioned the efficacy of capitalism. In this respect, the work created by Edward James could be described as surrealist architecture. From the perspective of rational, productive thought his buildings —which represent the work of 68 families over several decades and an enormous quantity of material and economic resources— serve absolutely no useful purpose. They are the mute testimony to his obsessions, a world of petrified dreams and nightmares. They materialize the image of the mediaeval castle occasionally inhabited by a phantom, archetype of the surrealist imaginary derived from the Gothic novel by André Breton and present in the cultured past of James himself, who was born in a neo-Gothic house in West Sussex designed by James Wyatt. And amid the disconcerting stylistic eclecticism there is also a reinterpretation of symbolic elements of Mexican culture.

Edward James commissioned the work as the ideas occurred to him and, without waiting for each one to be completed, overlapped project on project. The concrete was shaped in elaborate wooden formwork, crafted by hand. Here palaces, temples, pagodas and fountains inhabited by animals have their share here in the caprices of a fantasy world. James dreamed of sleeping in the house in the form of a whale, where the animals gathered. Following the labyrinthine course of his paths we come upon a circular door which gives access to a path flanked by serpents, the terrace of the tiger, the house of the parrot, series of incredibly slender bamboo-shaped columns that support nothing, doors that do not open, concrete extravaganzas in the form of immense trees, bridges such as the *Fleur de Lis*, columns with spiral stairs that lead to nowhere but the sky, plazas such as the Plaza Eduardo, gates such as that of St Peter and St Paul, a great library with no books, a highly unusual cinema, and other spaces designed to pay tribute to Henry Moore or to evoke Max Ernst.

All of the constructions, in full and harmonious integration with nature, are laid out around the estate's nine pools and the waterfall, the sound of which pervades the whole setting.

The House on the Rock Alex Jordan, Spring Green, Wisconsin, USA, 1945-
Alex Jordan was the second child of a family of wealthy cattle-ranchers and lan-
downers. His elder sister died before the age of three. Ingenious and a little spoiled,
even at school he stood out for his skill in drawing and the in making models. Tall
and well-built, he developed his exceptional physical strength playing American
football and working on construction sites. He dropped out of university during his
first year.
From an early age Alex Jordan felt a special devotion for the unique beauty of Deer
Shelter Rock. He used to climb to the top in order to contemplate the spectacular
view and dream of one day building his own private refuge there. Around 1945
Jordan started to lay down the first rocks of limestone, which he carried up on his
own back from the quarry of the estate itself. In this way he laid the foundations of
what we know today as The House on the Rock.
The solarium on the terrace offered fine views of the landscape to the north-west.
The living room had an enormous fireplace at one end, with the television and music
equipment on the upper level. The adjacent studio had a skylight that could be opened and closed automatically. The *Gallery of Bells* consisted of a narrow passage
giving access to the house. In the patio, with its garden, its ponds illuminated in
changing colours and its waterfall, Jordan and the woman he lived with regularly
organized special dinners and parties, to celebrate the 4th of July. The enthusiastic
comments of the guests, and above all the pressure put on him by his father, contri-
buted to his decision to open the house to the public in the summer of 1960. He had
made his dream reality in a physically tangible creative work, but in the last analysis it had been the money of his dominating father that had made the whole thing
possible, and he wanted a return on his investment.
From then on, Jordan supervised the continual extension of the house, which received an ever-increasing flow of visitors every summer. Using models, but with no
plans or drawings, he expressed the projects that took shape in his imagination,
which were duly carried out by a staff of up to 100 employees. His love of collec-
ting in general and in particular for doll's houses, carousels, weapons, automatons
and mechanical musical apparatus made it necessary to enlarge the house. Although
he hated travelling, he visited some of the suppliers of his workshops, in 65 different countries. Jordan went to great lengths to keep himself out of the media, in an
effort to conserve his anonymity. Complex and contradictory, he could be severe and
intimidating in his relations with his employees, whom he often dismissed at a
moment's notice, sometimes hiring them again the very next day. His relationship
with the visitors to the house was based on the sharing of the fantasy that his work
had generated.
The *Gate House* serves as the *foyer* of the main house, communicating with the
Gallery of Bells by way of a ramp at the level of the treetops. The *Water Garden* is
a paradise of aquatic plants, with a waterfall. The former solarium was converted
into an art gallery, with a three-storey bookshop. In 1962 Jordan installed a set of
enormous wind chimes. The *Mill House*, with its water wheel and the sound of three
violins and a piano, all operated pneumatically, is the entrance to *The Streets of
Yesterday*, a reconstruction of the atmosphere, the music and the houses of the
American West of the 1880s. In the *Blue Room* there is a symphony orchestra made
up of automated musicians. Jordan spent four and a half million dollars on the construction of the world's biggest carousel. The *Organ Room* is charged with an intense atmosphere; here Jordan sought to recreate the realms of the inferno and the sha-
dows. In 1985 he opened the *Infinity Room*, a project that had been taking shape in
his head over the course of 40 years. This room, 66 metres square, projects out into
the void, with large windows looking out on the landscape of the Wyoming Valley.
The subtle sloping away of the floor at the far end and the narrowing in of the walls
give the visitors walking around this room the sensation of finding themselves in an
endless space.

ST FRANCIS OF THE ROCK
CASA DEL ROCHE'
HOUSE ON THE ROCK
LITTLE SWITERLAND
DEER SHELTER
CASTLE IN THE SKY
ISLAND IN THE SKY
VILLA ST FRANCIS

General view of the exterior.
Below: various items from Alex
Jordan's notebook with possible
names for his paradise.

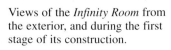
Views of the *Infinity Room* from the exterior, and during the first stage of its construction.

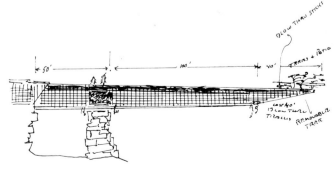

Previous page: the world's biggest carousel; the *Streets of Yesterday* decorated for Christmas; the *Blue Room* and the Mikado.

Tour Eben-Ezer Robert Garcet. Eben-Emael, Belgium, 1950-1964

Robert Garcet was born in 1912, into a heretical working-class socialist family. He first read the Apocalypse when he was still a young boy. At the age of twelve he was an ardent reader of Jules Verne. He received little formal education, a circumstance he regarded as very positive in that it allowed him to develop his imagination. He repeatedly declared that he wanted to discover something, although he was unable to define what this was. In due course he found his path thanks to his experience as a stonecutter, what he referred to as his "experience with silica". He made himself a geologist, palaeontologist, philosopher and writer. Convinced that it should be possible to undertake a reinterpretation of the Bible, he studied the past in order to obtain an understanding of the future. The mystery of the origin of things was the principal theme of his work.

Garcet published several books, of note among these being the *Heptameron*, in seven volumes, in which he declares that "... just as all things were created in six days and on the seventh day there was such repose that nothing more was talked of than Paradise and Man..." This work represents the revelation of the Apocalypse: "Heptameron" means seven days, seven periods, thus situating us in the mystery of the days of eternity.

The construction of the Tour Eben-Ezer and the *Heptameron* are two versions of the same story. The Tour Eben-Ezer is the *Heptameron* built in stone; the supporting structure for the messages of hope handed down by the prophets.

When he commenced construction of the tower he relied on the experience he had acquired in the reconstruction of his house, badly damaged by wartime shelling in 1940. Unable to obtain building materials, and being familiar with the techniques of masonry, he constructed his tower using stones from the neighbouring quarries and the volunteer labour of a number of his friends, whose names are inscribed on the walls. The Tour Eben-Ezer is a monument to fraternity.

Garcet began by reinforcing the old underground galleries which ran the length of his land. He then built a well, which was at the same time a geological section, and found water at a depth of 26 metres. The more he meditated on the words of the prophets, the more clearly the scheme of his tower took shape in his mind. Just as in the Apocalypse, the tower is constructed on the basis of the sacred number seven. The first floor rests on the Earth, the seventh is in Heaven.

In the *Heptameron*, the general scheme is a quadrangular prism which divides history into four different facets, four *phyla* that come together in the trunk of time. The Tour Eben-Ezer rises up out of deep, dark underground galleries. The cherubim support the tower with their wings outspread at its base, constituting a central pillar which on the ground floor carries the weight of the whole structure and culminates with the images of four creatures oriented towards the four winds of Heaven. The lion looks towards Jerusalem; the bull towards the north-west; the eagle towards the lands of Germany, Poland and Russia; and the lady, with her angel's wings and her beautiful gaze, is the sphinx who watches over the paths of the Earth.

For Robert Garcet, the subsoil is "a Bible of stones, it is the word of God. Silica is the stone of Man that speaks to us because for millions of years human beings and silica have been together, are of the same family, of the same race; unlike other stones, silica is alive, is formed in the womb of the Earth and petrifies with time. An archaeologist can be wrong, but not a stonecutter. My references are the mountains torn out page by page; they are 20,000 tons of worked silica".

Eben-Ezer is at once a place of revolution and of re-encounter. "In Eben-Ezer we live in anarchy, we are rebels, and with good reason, since the authorities are mere marionettes. We do not follow any religious tendency, any sect. Man should be the equal of other men, neither the tyrant of other men, nor the slave of other men. Without slaves and without masters".

Between reality and fiction, between madness and sanity, Garcet constructed a place between dream and reality that is a Utopia born of his understanding.

CLAUDE AND CLOVIS PREVOST, *Les Bâtisseurs de l'imaginaire*. Editions de l'Est, La Malgrange, 1990.

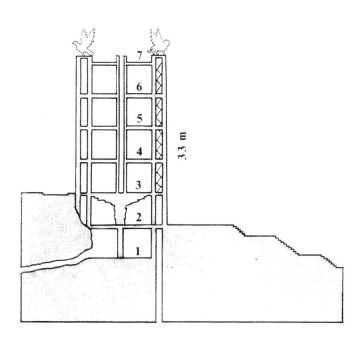

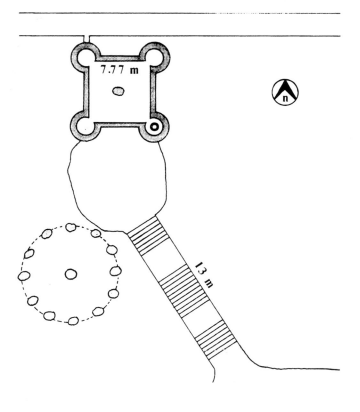

Plan of the complex, north-south section
and a period aerial photograph.

Robert Garcet on the roof in front of one of the four figures which crown the tower; in front of the Gospel According to St Matthew and, above, in front of some dinosaurs. Robert Garcet believed that the dinosaurs did not disappear from the Earth in the Cretaceous period, but survived in human form.

Views of the mermaids carved out
of the rock and of one of the
cherubs which support the tower
with their outspread wings. The
presence of imaginary elements
relates the world of legend to
that of everyday reality.

Casa O'Gorman Juan O'Gorman, El Pedregal de San Ángel, Mexico City, 1956-1961
Juan O'Gorman (1905-1982) was a well-known Mexican architect and painter. The construction of his own house is very often a difficult task for an architect, and this is the only house in this book created by and for an architect. Given the specific circumstances of the case, this is an exception to the rule, in that the building of this house not only marked a radical change of attitude in O'Gorman's career, but was also his last work.

O'Gorman's architecture is characterized by buildings of a rationalist nature which invoke logic and economy. Concerned with keeping down costs and completion times he proposed, for example, a rigorous standarization in the construction and remodelling of 25 schools. Faithful to the principles of functionalist architecture, he achieved international recognition with his Central Library on the University City campus in Mexico.

Although he had already confronted the theme of designing his own home as a young man, when he built himself a first, rationalist house, it was on this second occasion, at the age of 51, and looking now for what was to be his definitive home, that he changed his attitude: 'It now seems evident that the international style of the modern movement as applied to residential architecture has been exhausted... people are sick of the functional style called after the Bauhaus...' Concerned with the discovery of an authentically Mexican architecture and with the harmony between people and the Earth as a fundamental part of the human dwelling, O'Gorman found in popular architecture and in organicism the two pillars on which to base the design of his home. It was at this point that he took Wright as his inspiration, on account of his condition as the father of organic modern architeture and his ability to create a meso-American architecture (he regarded Taliesin as 'the finest modern project built this century').

Casa O'Gorman's in El Pedregal is built in the interior of a bed of volcanic lava, powerfully evoking a primitive cave. "All of the walls were constructed following in plan the form of the lava, with the double purpose of saving foundations and of obtaining forms suggested by the natural configuration of the rocky terrain that had been formed in the volcanic eruption. In order to obtain heat from the sun, he included a skylight to warm and illuminate the stone interior and the natural plants." This was very much in keeping with another of his observations: 'The popular architecture we have inherited from the past is an excellent source of inspiration to be applied to and utilized in the progress of the modern dwelling, because it possesses the necessary elements that are in harmony with the function and the climate of the region where they are situated... regional construction methods also achieve a natural simplicity which it is not only pleasant to live in, but is more economical'.

Pre-Colombian iconography, such as the figure of the rain god Tlaloc, is repeated in the stone mosaics that cover the walls of the house. The stones used created a palette of twelve colours, all of which were produced from local sedimentary volcanic rocks, with the exception of the blue, the only one that is not a natural mineral. The house makes very effective monumental use of masks derived from pre-Colombian, and in particular Mayan, architecture, while its situation reveals the architect's position in relation to native Mexican culture, as exemplified by the precise siting of a window especially to frame Popocatepetl and Ixtaccihualt, mountains considered sacred by the pre-Colombian culture.

However, this radical turnaround in O'Gorman's career was to go no further. After completing his house, at the age of 56 years he closed down his office and retired from architecture in order to devote his energies to painting, as he had done as a young man.

138

Exterior and interior views of the studio; view and detail of the dining room.

Exterior views; a serpent, regarded as a sacred being by the Mayan culture, slithers across one of the facades. The entrance to the living room is on the north facade. Above: detail of a skylight and mosaics on the roof.

Bottle Village Grandma Prisbrey. Simi Valley, California, USA, 1956-1981
Tressa Luella Schafer was born in 1896 in Easton, Minnesota. At the age of 15 she married a man 37 years older than she was. The mother of seven children, she left her husband in the late twenties, taking all her children with her, and earned a living as a waitress, pianist and singer. During the Second World War she worked on the assembly lines of the Boeing Corporation in Seattle, Washington.

In the early fifties she moved to Santa Susana, where she met her second husband, Al Prisbrey, whom she married in 1955, and they bought a plot of land in Santa Susana on which they settled down in a mobile home.

The space of the mobile home soon proved insufficient for Grandma Prisbrey, her grandchildren and her numerous collections of objects. The first of the houses she built on her land was the *Pencil House*, in order to house her constantly growing collection. Over the course of 25 years she was to construct a number of others, using objects and materials picked up on her daily visits to the local dump. Countless old bottles were bonded together with morter to form the walls of the little houses, as well as the wall that shelters the plot from the dust and smell of the neighbouring turkey farm. An even cheaper material than concrete blocks, the bottles compose translucent walls that light their fantasmagorical, unreal interiors.

The life of Grandma Prisbrey continued to be marked by hardship and loss. She buried six of her seven children, her husband and a fiancé. Nevertheless, she retained her vitalistic and positivist attitude and her exceptional ability to transform pain and sorrow into something more. Inside the houses there are numerous references to motherhood, to good wishes, to symbols of good luck and religious structures. The names of these spaces are themselves highly significant: *Rumpus Room*, *School House*, *Shell House*, *Blue Bottle House*, *Round House*, *Cleopatra's bedroom*, *Doll House*, *Little Chapel*, and so on.

The itinerary between the houses is lined with remarkable structures such as the *Leaning Tower of Pisa*, *Doll's Head Shrine*, *Wishing Well*, *Mosaic Card Figures*, *TV Tube Wall*, *Sanctuary Wall*, *Fountain*, etc...

Grandma Prisbrey gave a new function and meaning to the objects she found. She used television screens to construct a wall running the entire length of the plot. Car number plates and posters covered the walls. Dolls, hospital material, pieces of water pipe or car headlights were transformed into idiosyncratic sculptural structures such as the intravenous-drip-tube-fire-screen or the bath-bird-headlight.

Grandma Prisbrey took great pleasure in her Bottle Village, and shared it with whoever was prepared to pay the 25 cents for the guided tour she conducted in person, showing off and describing all of her houses, temples, sculptures and mosaics. And at the end of the tour she would play the piano and sing songs from the twenties.

Bottle Village found its first admirers among the local people, above all the children.

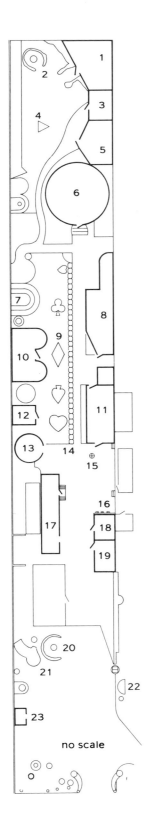

no scale

General plan of the complex
1 *School House*
2 Sanctuary
3 *Shell House*
4 *Blue Bottle House*
5 *Pencil House*
6 *Round House*
7 *Wishing Well*
8 *Cleopatra's bedroom*
9 *Mosaic Card Figures*
10 *Doll House*
11 *Rumpus Room*
12 Cabana
13 *Little Hut*
14 *TV Tube Wall*
15 *Leaning Tower of Pisa*
16 *Sanctuary Wall*
17 Caravan
18 *Bottle House*
19 *Shot House*
20 Thatched House
21 Fountain
22 *Doll's Head Planter*
23 *Little Chapel*

From top to bottom: detail of the perimeter wall; the *Blue Bottle Garden*, with the Sanctuary, the *School House* and the *Shell House* in the background; the *Doll House*.
Following page: *Doll's Head Planter*.

Rumpus Room.

The *Pencil House* and *Cleopatra's bedroom.*

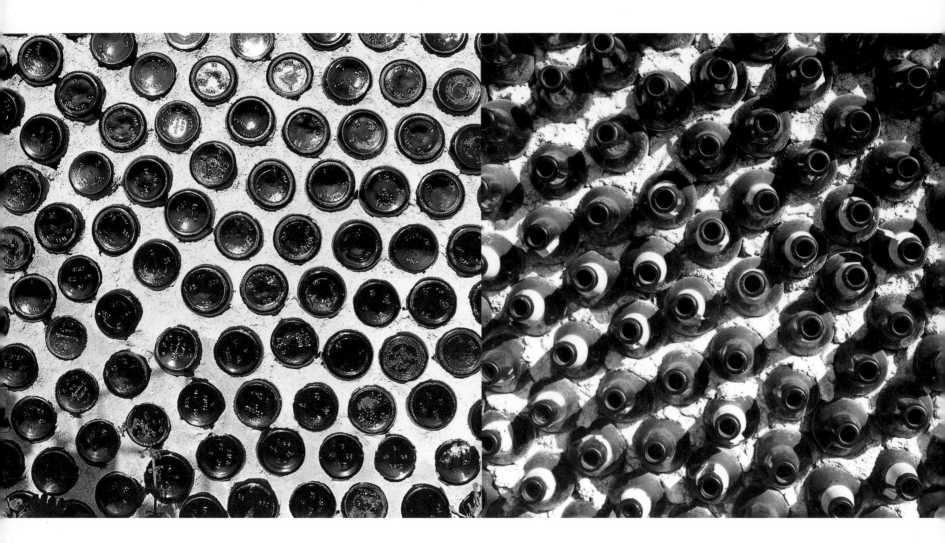

Following page: view of the entrance to the garden.

Paradise Garden Howard Finster, Pennville, Georgia, USA, 1961-

Situated some 160 kilometres north-west of Atlanta, Paradise Garden is a property of one and a half hectares on which the Reverend Howard Finster made tangible his own personal version of the Garden of Eden.

To quote Howard Finster: "I have never had a traffic fine in spite of having driven the length and breadth of the country, I have never got drunk, I have never robbed, or been arrested, or been in prison. I have never stolen anything from anybody. I have never been unfaithful to my wife. I play the banjo. I have paid all my debts. I have been a preacher in various churches and, as well as that, I have built bicycles for poor folks, and I have made clocks to sell and make a little money to be able to live and buy the land to build this garden. I have also repaired lawnmowers. I have learned a lot from my body. I have had visions observing my stomach and my brain. I am a man of vision. I am a second Noah who has come to the world before it is too late. In a way, I am more successful than Noah. He preached to the world but he didn't save anyone. As far as I know, the only ones who were saved were his family. I am the second Noah and I am leading my family and the whole world towards God. The people are being saved because they write to me to tell me they have seen my work and have heard my message, that they have found in it the meaning of life and that now they are happy thanks to its existence".[1]

Finster's artistic achievement is a direct consequence of his mission of serving God. His work is the message he has to transmit. His art is thus a sacred art. It is a work embodied in his garden, which is transformed into a place of worship and spreads his message to the outside world. The garden is open to visitors free of charge. In it they can walk wherever they choose, eat the fruits and berries provided by nature, even camp out.

In 1961 Finster bought a plot of land for $1,000, on which to construct his vision of Eden. This was a piece of waste ground situated in a wet zone so muddy that it was necessary to wear high boots. As he had neither money nor materials, he used his own works of art and the multitude of objects he found in the neighbouring scrapyards and demolition sites and restored in order to materialize his dream. Finster is a kind of artist-handyman who has described himself as God's Junkman. For him, nothing is waste, everything is capable of being reused. Matter is in continual metamorphosis.

There are a number of different buildings in the garden, of note among these being *The Front Molding*, *The Folk Art Chapel*, *The Pump House*, *The Studio*, *The Tin Chapel* and *The Bible House*.

The entrance is on the south side, where a path leads to a small building which was originally Finster's own house and then his studio. Behind the house is *The Folk Art Chapel*, which is the largest construction in the garden. This is a circular tower five storeys high with a spire on top, which contains an assortment of books, press cuttings, photographs and gifts that Finster has received over the years from friends and visitors.

The topography of the terrain makes access difficult for people with diminished mobility, with the result that Finster decided to build a covered bridge, *The Rolling Chair Ramp*. This is at the same time an art gallery, open to any artist who wishes to put up their work and telephone number.

Next to the bridge is a large sculpture composed of bicycles and lawnmowers tied together with wires, *The Bicycle Tower*.

Finster's pictures, most of them painted on wood and cut out around the silhouette of the person portrayed, are all over the garden, hanging from every possible support and populating it with personalities.

Similarly, there are written messages on all sides. Finster's garden is literally covered with verses from the Bible. The message is always the same —'Turn away from Satan and the devil and follow Jesus and the way of the Cross'— although expressed in different ways.

[1]. ROBERT PEACOCK WITH ANNIBEL JENKINS, *Paradise Garden*, Chronicle Books, San Francisco, 1996

Detail of one of the concrete
'mountains'. Above: view of the
Pump House.

The *Bible House* with numerous messages made up of cut-out letters tacked into place. Above: detail of the *Folk Art Chapel.*

The *Bird's cage* is made of
recycled refrigerator trays.
Above: views of *Smashed car*
(sculpture-message for drunk
drivers); the studio and the
Bicycle Tower. Facing page:
the *Tin Chapel*.

Little Sparta Ian Hamilton Finlay, Dunsyre, Strathclyde Region, Scotland, 1966-
Ian Hamilton Finlay was born in Nassau (Bahamas) in 1925. He studied at Glasgow School of Art for a short period in the early forties, before doing his military service, and from that time on has been an autodidact. Nowadays his work is well-known and very highly thought of. He is one of the leading representatives of 'concrete poetry', producing texts of extreme astringency which transmit a very powerful charge of signification with the minimum of verbal material. In 1945 he settled in the Orkney Islands, where he farmed and kept sheep, and it was there that he started to write short narratives. He moved to Edinburgh towards the end of the fifties and wrote his first poems. He subsequently lived in Easter Ross, north of Inverness, with his first wife and baby, in a small house without running water, until in 1966 they decided to move to a piece of land belonging to his wife's father, which is now the site of Little Sparta. On this land there was a farm, long abandoned and unoccupied, which they set about refurbishing. The farm, to which they initially gave the name of Stonypath, was in a remote, wet area good for little other than grazing sheep. From that time on Ian Hamilton Finlay has worked year in, year out on the creation of this unique place, where he has established his home and his studio, and has transformed it into one of his most important works. Little Sparta is not to be regarded as the fantasy of some naïf, but is quite possibly the greatest achievement of Finlay's entire artistic career.

Here the poet has built ponds, planted bushes and hedges, and has placed hundreds of inscriptions and objects, thus converting the place into a metaphor of the world we live in, a battle against the absolute secularization of contemporary culture, in which he sees everything reduced to the level of banality. Little Sparta is a reinterpretation of neoclassicism and, in consequence, of revolution. The whole repertoire of the history of European culture is utilized as a source of references. Hamilton Finlay sees in the French Revolution, in particular, a perfect example of the dialectics between culture and nature, between reason and terror. He has superimposed his own layer of meaning on the landscape, deploying poems, aphorisms and quotations chiselled in stone all around the garden.

The construction of the garden has been a gradual process, undertaken on the basis of small individual areas, with each area having its own particular artefact. However, the work is not the artefact but the composition as a whole. Each piece asks to be appreciated in its context. The work is the composition, and the composition is more than an isolated individual object, in the same way that a letter takes on meaning in a word, a word in a sentence and a sentence in a text, and even the text only becomes properly intelligible when considered in context. The garden of Little Sparta is itself a poem. Finlay has engaged here with the issue of how to put language into nature. He began with the sundials, moving on to the benches, the plaques on the trees, then the bases of the columns for the trees, etc., effectively creating a language for the purpose. For the material realization of this language he has worked in collaboration with a number of different sculptors.

Just as in the historical garden, here there is a constant confrontation between the world of pastoral and the presence of military symbolism. Submarines, hand-grenades, aircraft carriers, etc., caustically embody this contemporary interpretation of Arcadia.

The cow-shed has been converted into a gallery for displaying the works that could not be exhibited in the house or outside in the garden. Strongly opposed to the connotations of the word "gallery" (money, tourism), Hamilton Finlay decided to call his a *Garden Temple*, thus changing the nature of the place. This is not a space designed to contain the works, but one in which the works on show define the space.

Entrance to the main garden.
Facing page: *Garden Temple*.

Grotto of Aeneas and Dido; a series of moss-covered stones form a little path which crosses the middle pond, and one of the sundials (Hamilton Finlay regards the sundial as a perfect union between philosophy, poetry and craft skills). Below: *Hypothetical entrance door to the Academy of Mars* (1991 with David Edwick) and *Apollon terroriste*, (1988 with Alexander Stoddart). In Little Sparta, the figure of Apollo often stands for the French revolutionary Saint-Just.

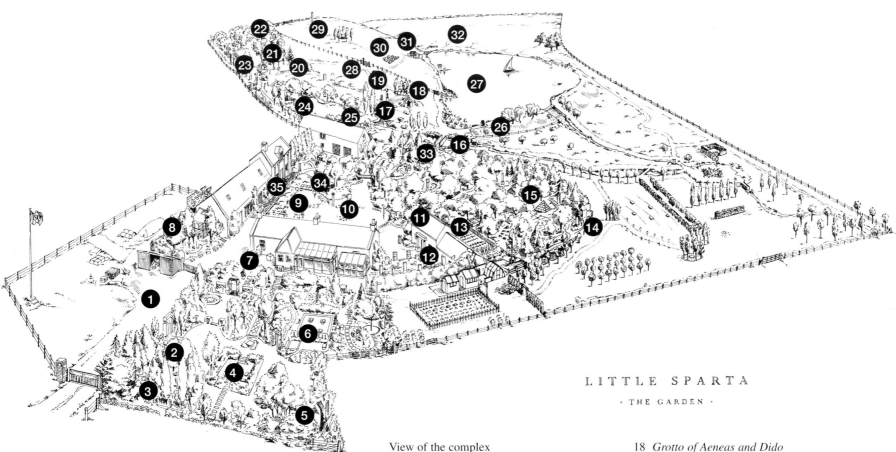

LITTLE SPARTA

· THE GARDEN ·

View of the complex
1   *Entrance to the main garden*
2   *Roman garden*
3   *Henry Vaughan walk*
4   *Sunken garden*
5   *Mare Nostrum*
6   *Raspberry camouflage*
7   *Sundials (Fragments / Fragrance)*
8   *Julie's garden*
9   *'Das grosse Rasentück'*
10 *Temple pond*
11 *Temple of Philemon and Baucis*
12 *Lararium*
13 *Kitchen garden (Epicurean)*
14 *Stone with inscription 'Pacific Air'*
15 *C. D. Friedrich pyramid*
16 *Claudius' bridge*
17 *'Xaipe' after J. C. Reinhart*

18 *Grotto of Aeneas and Dido*
19 *Hypothetical entrance door to the Academy of Mars*
20 *Hillside pantheon*
21 *Silver Cloud*
22 *Virgil's spring*
23 *Upper pond*
24 *Middle pond*
25 *Apollo and Daphne*
26 *Nuclear navigation*
27 *Lochan Eck*
28 *Hegel 's steps*
29 *Half-way inscription*
30 *'The Present Order...'*
31 *Laugier's hut*
32 *The column of Saint-Just*
33 *'O Tannenbaum'*
34 *Tristram's sail (sundial)*
35 *The Garden Temple*

'The Present Order...' is a work
based on a quote from the French
revolutionary Louis-Antoine Saint-
Just (1767-1794), produced in
1983 with Nicholas Sloan.
Above: view of *Claudius' bridge*

View of the model of a boat in the living room and detail of the bath. Above: view of the fireplace in the living room and details of the wardrobe and the bathroom door. Below: other model boats floating on the pond or in 'dry dock' on the living room mantelpiece, creating a series of miniature landscapes.

À MARAT

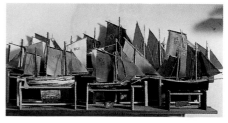

Villa Spies Simon Spies, Töro, Sweden, 1967-1969

Even as a small boy Simon Spies stood out as a good student and for his initiative (he set up his own business selling vegetables or "energy stimulants" which were in fact nothing more than tablets of lime). Modest ventures, but not lacking in the visionary intuition or the necessary ambition that would enable this boy from a humble family to make himself the boss of his own empire. After studying for two university degrees at the same time, he found occasional work as a tour guide in Mallorca for a number of seasons. Once he had come to understand how the tourist sector functioned, and the opportunities for low-cost tourism represented by the vacant accommodation in winter, in 1965 he launched his own tour business with an advert in the newspapers. A good part of the company's publicity was generated by the sensationalist news stories about its extravagant founder, who liked to wear leopardskin suits, drive a different car each day and appear in the company of a different young woman every morning; it came to be known as the 'Spies Circus'.

Simon Spies thought of the entire population as potential tourists with expectations for their leisure time that went beyond the provision of the traditional hotel room with television and continental breakfast. In places with beaches and palm trees, but little or nothing in the way of cultural life, he set out to offer 'bubbles of pleasure'.

In 1967, the Spies company organized a competition to make this concept a reality. The winning project, by the Swedish architect Staffan Berglund, was never built. Nevertheless, the way Simon Spies had imagined his own personal habitat was not much different from Berglund's proposal, and he commissioned the architect to design and build it. The Villa Spies was completed in 1969, the year that Neil Armstrong set foot on the moon. The silhouette, in the form of a hemispherical shell at the base with a lightweight vaulted roof, suggests a flying saucer resting on the surface of the Earth, next to the sea. The base of the volume, circular in plan and built of *in situ* concrete, gradually opens up as it rises. In addition to aesthetic and climatic reasons, the inclination of the windows was determined by the transmission of the torsion produced by wind loads and the weight of the vaulted roof, which was constructed from prefabricated plastic radial sections ten centimetres thick.

The villa is relatively small, and its concentrated form accentuates this impression. Ingenious experimental solutions have been adopted to stimulate the senses and to meet the changing requirements of the user in different situations. Atmospheric control is here converted into control of the space, in the sense that it defines the perception of this.

In the open space of the upper floor there are two containers with folding walls which open up to reveal the main bedroom and the kitchen-cum-office. The long semicircular sofa defines the main meeting space, in which all of the people taking part can see and hear one another. The slightly lower level here steps down towards the surface of the heated swimming pool and at the same time avoids obstructing the views. Whenever required, by pressing a button a table and six chairs rise up from the centre of the space to create a dining room for six. In this raised position on the upper part of the mechanism there are fine views of the surrounding landscape. The food is served from the kitchen on the lower floor.

A central control panel serves to adjust the parameters of light, sound, image reproduction and relation with the exterior, thus creating the desired atmosphere. The circle suspended from the middle of the ceiling supports spotlights and slide projectors. Landscapes, works of art and photographs of all kinds can be screened on walls, windows, floor and ceiling. The intensity and colour of the light can be varied to create different effects, including the simulation of natural light. The telephone, television and music system can be beamed to radio headphones. Concealed alongside the air-conditioning outlets (which resemble aircraft engines) on the facade are loudspeakers which, thanks to their radial, equidistant position, make it possible to produce effects of movement of the sound source. The natural light can be controlled by two different kinds of motorized screens.

Staffan Berglund's Villa Spies, text by Mikael Askergren, Eriksson & Ronnefalk Förlag, Stockholm, 1996.

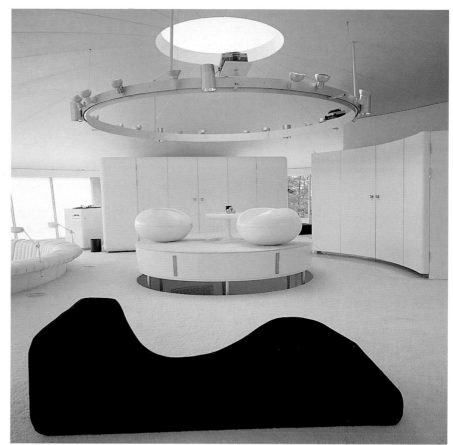
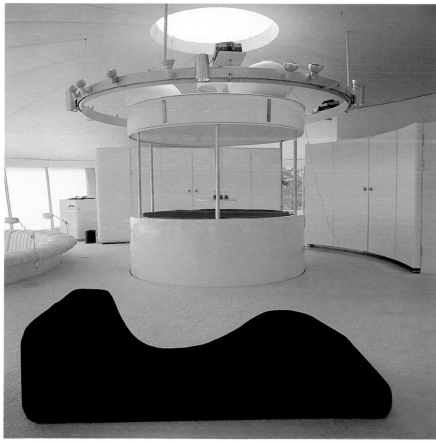

The central space is made up of a series of furniture-containers accommodating the bedroom and the kitchen, which can be opened up or closed completely to transform the space. On the following page: the white surface of the closed furniture-containers serves as a backdrop on which to project images to create different atmospheres.

Casa-estudio César Manrique César Manrique, Taro de Tahíche, Lanzarote, Spain, 1968-1992

The remarkable volcanic landscape of Lanzarote is characterized by the more than 300 craters which irrupt from its smooth topography, by the exhuberant tropical vegetation covering a third of its surface, and by the tonal quaity of the volcanic pumice scattered with sand carried in on the wind from the nearby African desert. César Manrique Cabrera was born in the island's capital, Arrecife, in 1919.

After the end of the Spanish Civil War, Manrique enrolled in the Universidad de la Laguna to study technical architecture. In 1945, two years later, he moved to Madrid, after receiving a grant to study painting at the Escuela de Bellas Artes de San Fernando. In 1954 he was one of the figures actively promoting pioneering exhibitions of abstract art and the renovation of postwar Spanish art. His non-figurative art explored material techniques in order to achieve the maximum of expressive power with the minimum of resources. Following the death of his companion in 1963, and having achieved international recognition as an artist, he moved to New York, where he was in direct contact with American abstract expressionism.

When he returned to Lanzarote in 1968, Manrique set out to develop a Utopia based on a commitment to protecting the natural environment of the island, conceived as a space of artistic and cultural integration. He worked in the fields of painting, sculpture, murals, architecture and urbanism, as well as the promotion of cultural centres and museums.

The construction of his house-cum-studio brought back to Manrique the special sensations of his childhood, instilling in him once again the magic of the volcanic fissures, the torrent of lava produced by the volcanic eruptions. This is the materialization of a dreamed-of fantasy, nourished by his nostalgia for the island in the years he spent living abroad.

Seen from a distance, the Casa-estudio César Manrique can be confused with any of the whitewashed cubic volumes scattered over the plains and slopes of the island, with the volcanic peaks of Maneje and Tahíche as a backdrop. However, the interior of the house is a metaphor of Manrique's personal and artistic experiences, a landscape of volcanic soil and exotic vegetation framed in an architectural symbiosis of the vernacular, resulting from an abstract rereading of the island's traditional architecture and the modern. Like some living organism, the house grew from its original nucleus, responding to the needs, the spontaneity and the intuition of César Manrique.

A cavernous labyrinth of tubes, partially tunnelled through the hard volcanic rock, connects the five bubbles and the swimming pool with its organic forms, an oasis in the desert of lava. These bubbles, named after the colour of the lower part, are spaces of an almost "found" architecture, where the plasticity of the characteristic geology of their walls is accentuated here and there by the overhead light. Manrique transformed these interiors into habitable spaces, cladding the floors and the lower part of the walls and introducing pieces of design and *objets trouvés*, of intense colours. These are magic spaces, contemplative in character, where the nature of the earth and the vegetation engage in a dialogue with the artificiality of the intervention, linked together in a relationship of reciprocal enhancement.

The beauty of the natural elements, their brutal materiality and the richness of their forms and textures are exalted and enhanced by Manrique's potent and thoroughly thought-out plastic and tectonic interventions. And vice versa.

The visually continuous space of the upper floor of the house-cum-studio achieves a fluid fusion between interior and exterior, between architecture and nature. The different volumes and their spaces grow from one another in sequence and potentiate the diagonal circulation routes. The construction combines traditional materials and techniques with elements of the international style, such as the large glazed openings which give access to the terraces or the skylights which illuminate the interior. César Manrique was killed in 1992 in a traffic accident near the house-cum-studio which only a few months before he had presented as the headquarters of the foundation which bears his name.

Main entrance.

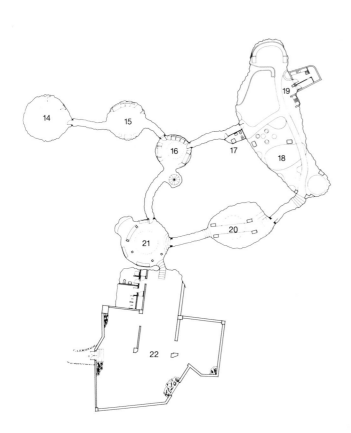

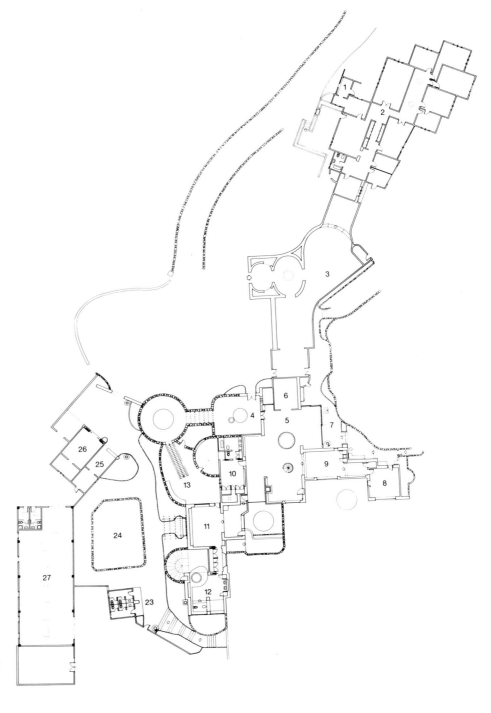

Lower floor and first floor plans
1 Ticket-office
2 Former servants' accommodation (now offices)
3 Temporary exhibitions gallery (former terrace)
4 Entrance courtyard
5 Former living room (now the paintings collection)
6 Former kitchen (now the collection of graphic work)
7 Terrace
8 Former 'cuarto de la jaima' (now the collection of Manrique's projects for intervention in the environment)
9 Former corridor (now the paintings collection)
10 Former guests' room
11 Manrique's former bedroom
12 Bathroom
13 Descent to the volcanic bubbles
14 Avocado bubble (fountain bubble)
15 White bubble
16 Red bubble
17 Bathroom
18 Swimming pool
19 Kitchen
20 Black bubble
21 Yellow bubble
22 The artist's former workshop and subsequent extension (now collection of Manrique's paintings)
23 Public toilets
24 Garden
25 Former garage (now café-bar)
26 Former garage (now shop and bookshop)
27 Storeroom

South facade of the temporary exhibitions gallery and terrace on the south side of the house. Above: interior views of the former living room. On the facing page: view of the swimming pool with the water spout on the volcanic wall.

White bubble.

Red bubble. On the following page, above: detail of the yellow bubble.

Villa Principe Vittorio Emmanuele di Savoia Prince Vittorio Emmanuele di Savoia, Island of Cavallo, Italy. 1980

In the late seventies, on one of their many trips, Princess Marina Doria and Prince Vittorio Emmanuele of Savoia fell in love with the enchanting island of Cavallo, situated between Corsica and Sardinia, in the Strait of Bonifacio. Since then, they have never ceased in the task of constructing their house on this little piece of paradise.

In 1980, they brought in the architect Savin Couelle to help realize their dream. This Frenchman of Spanish origin, whose father had pioneered the development of Porto Cervo on Sardinia, was equally taken with the place and undertook to construct the house in such a way as to conserve the equilibrium of the landscape. The integration of the human habitat into the environment set out to respond to two apparently contradictory objectives: the house was to be virtually invisible from the exterior while the interior was to open up to the surrounding landscape. On the basis of these premisses, the solution adopted was to construct a camouflaged house. For this purpose a series of floor slabs were constructed, transitable on different levels, and these were clad with local Bastia stone of the same tone as that found in the area. The interiors were then excavated to produce habitable spaces, creating an artificial grotto.

The distribution of the different rooms was determined by the presence of a number of enormous rocks, polished by erosion. These strange rocks scattered across the terrain have been respected and left in their magical place, and have become protagonists of the interior space. According to their situation, they assume different functions, such as the head for a bed, a shelf for objects, a plinth for the television, etc. Their organic geometries ensure the harmony of all of the interventions. Doors and windows have capriciously rounded forms. Where furniture is required, unworked natural stones are used, as in the outdoor dining area, with its stone table and its benches characteristically covered with cushions patterned with leopardskin fabric. Some of the doors are clad with tin or leather. The primitive cave and the life of the troglodyte are the mythical referents for this artificial grotto.

On the exterior, these gigantic rocks constitute a protective screen, while forming a natural harbour in front of the house which was used for disembarking the construction materials transported here by sea.

Plan.

Following page: View from the
sea, with the little natural harbour
in front of the house, and view of
the transitable roof.

Das Paradies Cornelius Kolig, Vorderberg, Gailtal, Carinthia, Austria, 1980-

Das Paradies consists of a series of buildings, constructed one at a time, whose meaning derives from their place in the complex as a whole (still unfinished), as this is developed over a number of years. The concept of paradise, its historical evolution, the stratification of its semantic signification and the possibility of utilizing this as a theme for the creation of art have been synthesized by the artist Cornelius Kolig in a group of buildings. Starting out from the architectural typologies of paradises already designed or built in different parts of the world, here he has given the concept a new and personal form and function.

The external appearance of Das Paradies is that of a construction produced from basic materials (wood, concrete blocks, aluminium, etc.). Access to the complex is by way of a narrow entrance on the north side of the perimeter wall. As an additional element of protection there is a screen of bushes and trees. The inscription ATA, the brand name of a well-known detergent, is here an allusion to the act of purification required in order to enter the precinct.

Once in the interior, the height of the perimeter means that the only views are those seen by looking upwards, towards the natural environment of the mountains and the sky.

The complex is based on the floor plan of a basilica, with two towers rising up above it with antennae on top, symbols of the electronic eyes and ears of this paradise. The additional buildings and chapels have a particular significance, and in some cases make reference to Christian iconography. The central space is an atrium in which a great abundance of flowers, climbing plants and shrubs are laid out in co-ordinated sequences of colours and essences according to the season of the year. These are left to grow in so far as they do not upset the aesthetic and intellectual order of the complex, and thus embody a domestication of nature. This isolated zone is dedicated to silence, peace and reflection. Around it, various sound installations allude to human nature , in particular to the processes and products of excretion. For example, there is a series of small chapels dedicated to urine, excrement and semen, which constitute the *Pantheon*. For example, the zone devoted to urine consists of a urinal for women covered by a yellow canopy, from which the urine is conducted down a sloping concrete surface to a nearby meadow. The *Observatory of the Sun* is a dark, cylindrical body with a hemispherical roof, where a narrow slit allows the entry of the sun's rays during certain hours of the day. Here the solar rays are reflected in the crystalline surface of the water, producing reflections on the concave wall which are only distorted when, from the bridge above, the visitor urinates in the water, creating concentric ripples which modify the reflection and the silence.

This is an itinerary that is dangerous for the senses, a labyrinth of emotions and an instrument for the analysis of the human body that can provoke irritation. The complex as a whole serves as a setting for the exhibition of Kolig's installations and other works of art related to the body, sexuality, death and our relationship with the nature. While in the mythological paradise the body disappears and only the spirit achieves heavenly rest, in his Das Paradies, Kolig treats the human body as a mechanism.

The sense of spatial isolation, that illusory characteristic of paradise, is only mediated by the views of the stables nearby, shown on video screens.

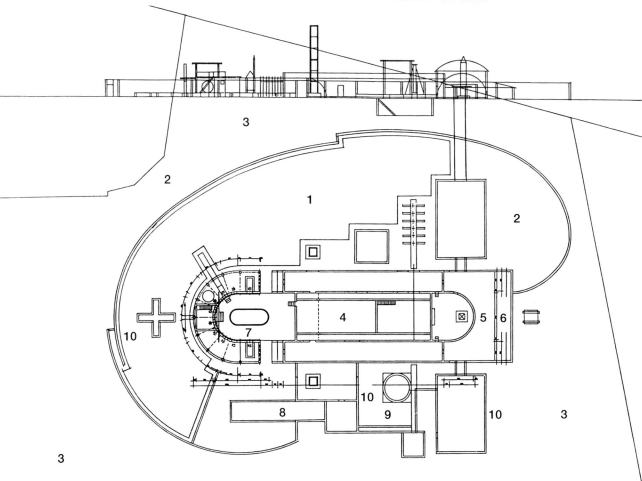

Plan and elevation
1 Field
2 Fruit trees
3 Meadow
4 Flowering plants
5 Vines
6 Wild mushrooms
7 Trees
8 Water
9 Vegetable plots
10 Climbing plants

Six concrete columns dedicated to six saints and view of the wall with the word 'ATA', the name of a well-known brand of detergent, symbolizing the act of purification required before entering Paradise.

Views of various installations.
The human body and its processes
of excretion and their products are
the central theme of Cornelius
Kolig's work.